Anna Friebe · Florian Haug
Editors

Robotic Sailing 2015

Proceedings of the 8th International Robotic Sailing Conference

Springer

Editors
Anna Friebe
Åland University of Applied Sciences
Mariehamn
Finland

Florian Haug
Åland University of Applied Sciences
Mariehamn
Finland

ISBN 978-3-319-23334-5 ISBN 978-3-319-23335-2 (eBook)
DOI 10.1007/978-3-319-23335-2

Library of Congress Control Number: 2015946769

Springer Cham Heidelberg New York Dordrecht London

Printed on acid-free paper

Springer International Publishing AG Switzerland is part of Springer Science+Business Media (www.springer.com)

Preface

Within living memory, mankind has been taking advantage of wind and water for power generation. Sailing ships and boats have used the wind as propelling force for centuries and contributed crucially to the discovery of new lands and to global trade. Maneuvering a sailing boat from a given starting point to a defined destination under changing wind, weather, and sea conditions is a highly nontrivial task; the skipper needs to adapt to changing parameters and optimize the boat's course during the entire journey. Moreover, sailors often have to sail through narrow passages or within areas with obstacles and dense traffic, simultaneously being aware of and following collision preventing regulations.

While a human being masters many of these complicated steering- and regulation problems through experience and intuition, it turns out to be a challenging task for an autonomous, unmanned sailing device. To maneuver a sailboat safely through changing wind and sea conditions without any remote control or human assistance is a complex and multidisciplinary mission, which involves boat designers, naval architects, electrical engineers, and computer scientists. Many of the challenges in building a truly autonomous sailboat are still unsolved and have stimulated a variety of groups all over the world. Intensive research and development is done within platform optimization, route and stability planning, collision avoidance, power management, and boat control.

Reliable autonomous robotic sailing devices have a huge potential and can be used for a broad variety of practical applications. Self-powered sailing devices are perfectly suitable for monitoring or collecting oceanographic, environmental, meteorological, or hydrographic data. Robotic sailboats can even contribute to traffic monitoring, boarder surveillance and security, assistance and rescue, harbor safety, and carbon neutral transportation.

These proceedings present the cutting edge developments within a broad field related to robotic sailing. The contributions were presented during the 8th International Robotic Sailing Conference organized by Åland University of Applied Sciences and took place in Mariehamn, Finland. In conjunction with the IRSC the World Robotic Sailing Championship 2015 (WRSC 2015) was held. The WRSC 2015 consists of a series of short-distance races, navigation, and autonomy challenges. The competition proposes tasks such as station keeping, speed in different conditions, accuracy, area scanning, and data collection.

The IRSC/WRSC 2015 is the 8th edition of the event. Previous IRSC/WRSC events have taken place in Ireland (2014), France (2013), Wales (2012), Germany (2011), Canada (2010), Portugal (2009), and Austria (2008).

The proceedings are divided into three parts. The first part is devoted to different hardware solutions for autonomous sailing platforms, including design, construction, test sailing, and applications of different devices. The following part presents ideas related to mission planning, localization, and obstacle avoidance. The third part is dedicated to the design, modeling, and evaluation of sensors and controllers. The allocation of the papers to the different parts might seem somewhat arbitrary, since some of the papers treat several of the above-mentioned aspects. These papers were placed in the category where they seemed to fit best.

The editors would like to thank all of the authors, the Program Committee, all the sponsors, partners, and other people who made it possible to host the IRSC/WRSC 2015 in Mariehamn, Åland.

Mariehamn
July 2015

Anna Friebe
Florian Haug

Organization Committee

General Chair

Anna Friebe, Åland University of Applied Sciences, Finland

Proceedings

Florian Haug, Åland University of Applied Sciences, Finland
Anna Friebe, Åland University of Applied Sciences, Finland

Co-organizers

Emmy Karlsson, Åland University of Applied Sciences, Finland
Sven Schauman, Åland University of Applied Sciences, Finland
Ronny Eriksson, Åland University of Applied Sciences, Finland
Therese Sjöblom, Åland University of Applied Sciences, Finland
Lena Nyman-Wiklund, Åland University of Applied Sciences, Finland

Program Committee

José Carlos Alves, INESC TEC (formerly INESC Porto) and Faculty of Engineering, University of Porto, Portugal
Taylor Barton, Sea Scope, USA
Ole Blaurock, Fachhochschule Lübeck, Germany
Jorge Cabrera-Gámez, Universidad de Las Palmas de Gran Canaria, Spain

Benoit Clement, École nationale supérieure de techniques avancées (ENSTA) Bretagne, France
Vincent Creuze, LIRMM, University Montpellier 2—CNRS, France
Nuno A. Cruz, Universidade do Porto, Portugal
Kjell Dahl, Åland University of Applied Sciences, Finland
Anna Friebe, Åland University of Applied Sciences, Finland
Oren Gal, Technion, Israel
Florian Haug, Åland University of Applied Sciences, Finland
Luc Jaulin, École nationale supérieure de techniques avancées (ENSTA) Bretagne, France
Fabrice Le Bars, École nationale supérieure de techniques avancées (ENSTA) Bretagne, France
Erik Maehle, University of Lübeck, Germany
Benedita Malheiro, Instituto Superior de Engenharia do Porto, Portugal
Paul Miller, US Naval Academy, USA
Fearghal Morgan, National University of Ireland, Galway, Ireland
Cedric Pradalier, GeorgiaTech Lorraine, France
Kostia Roncin, École nationale supérieure de techniques avancées (ENSTA) Bretagne, France
Colin Sauze, Aberystwyth University, Wales, UK
Alexander Schlaefer, Hamburg University of Technology, Germany
Michael Schukat, National University of Ireland, Galway, Ireland
Roland Stelzer, Happylab GmbH, Austria
Dermot Tynan, Hewlett Packard, Ireland
Diedrich Wolter, University of Bremen, Germany

Contents

Part I
Sailboat Platforms and Applications

A-TIRMA G2: An Oceanic Autonomous Sailboat

Antonio C. Domínguez-Brito, Bernardino Valle-Fernández, Jorge Cabrera-Gámez, Angel Ramos-de-Miguel and Juan C. García

Abstract This paper describes a new design of a 2 meter LOA (Length Over All) autonomous sailboat conceived for sailing in an ample set of weather conditions. The design has been focused on robustness and on achieving some degree of redundancy on critical components like sails and rudder. Accordingly, it is equipped with two light-weight carbon fiber wing sails and two slanted rudders protected by skegs. Its stability curve is fully positive, so she is capable of recovering autonomously from capsizing.

1 Introduction

The design of an unmanned sail boat, specifically conceived for sailing in an ample set of weather conditions, implies to take into account many particular factors to make it adaptable and flexible enough for distinct situations without human intervention. This could be done using complex and costly electromechanical systems,

A.C. Domínguez-Brito (✉) · J. Cabrera-Gámez
Instituto Universitario SIANI (www.siani.es) and Departamento de Informática Y Sistemas (www.dis.ulpgc.es), Universidad de Las Palmas de Gran Canaria (www.ulpgc.es), Las Palmas de Gran Canaria, Spain
e-mail: adominguez@iusiani.ulpgc.es

J. Cabrera-Gámez
e-mail: jcabrera@dis.ulpgc.es

A.C. Domínguez-Brito · B. Valle-Fernández · J. Cabrera-Gámez · A. Ramos-de-Miguel · J.C. García
Asociación Para El Desarrollo de Sistemas Marinos Autónomos - ADSMA (www.adsma.org), Las Palmas de Gran Canaria, Spain

B. Valle-Fernández
e-mail: bvallef64@gmail.com

A. Ramos-de-Miguel
e-mail: aramos.gcc@gmail.com

J.C. García
e-mail: juancarlos.feroher@gmail.com

A. Friebe and F. Haug (eds.), *Robotic Sailing 2015*,
DOI 10.1007/978-3-319-23335-2_1

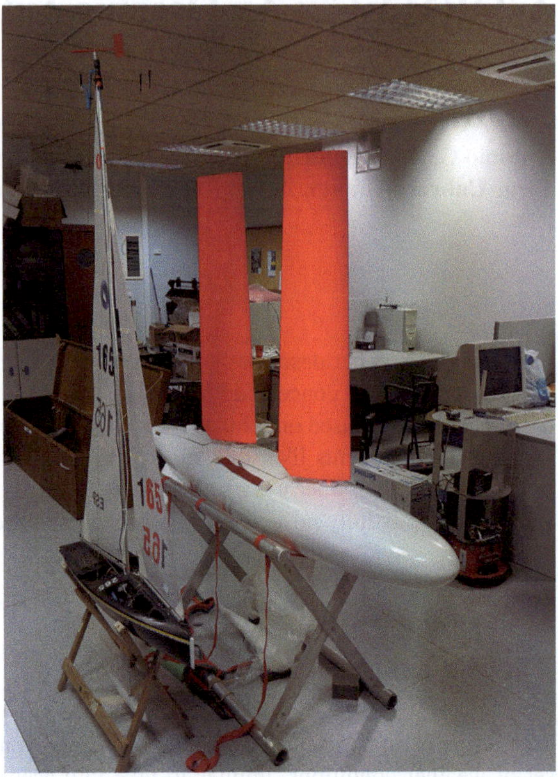

Fig. 1 A-Tirma G1 and G2

simulating the maneuvers typically performed by a crew. All this to keep the necessary balance between sail surfaces and wind speed. But, in the typical conditions found at sea, the probability of something going wrong is really very high. To minimize problems or the effect of complete breakdown of some elements, we might look for a flexible design and duplicate all elements and systems on board, in order to keep the boat sailing, even in precarious conditions. In addition, there are other problems to take into account, like avoiding marine traffic, or encountering floating debris or algae which might diminish the navigation capacities of the sailboat.

On the other hand, when designing an unmanned sailboat, there are some meaningful requirements present in a traditional vessel which are not necessary to fulfill, mainly those aimed to guarantee the habitability and maneuverability of the boat for the crew.

In this paper we will describe the design principles that have guided the design of the next generation of our boat A-Tirma [1, 2]: A-Tirma G2 (Generation 2). A-Tirma, on its first generation [1] was based on a commercial one-meter low cost RC boat (in Fig. 1 we can see both sailboats side by side in our laboratory). The motivation behind this initial approach was to get an affordable open experimental platform which could serve as test bed for the development of navigation algorithms for sailboats. A-Tirma G2 design tries to surpass some of the limitiations of the previous prototype, specially a better behavior in harsh conditions. On the other side, we got to a point with the fist prototype where there was no more room for payload on it. A-Tirma G2 is bigger in size having a length of 2 meters, allowing for loading more equipment on board. And in relation to its physical design, it is a new design developed from the scratch, where not having a deck and being provided with two wing sails are its main features.

2 Vessel's Description

The length of a vessel, not devised for gliding, conditions strongly the speed it can reach, as the speed is a function of the resistance by wave formation, and inversely proportional to the vessel's length. A greater length improves sailboat's navigation behavior, being less influenced by the impacts and thrust waves produce. Being said that, the length chosen for our design is mainly restricted by the practical aspects of having a prototype at a lesser scale, more manageable and economical. In Table 1 we can see the main dimensions of A-Tirma G2, and in Table 2 the more important properties related to its design. In Fig. 2, a perspective of the design is shown, and in Fig. 3 a profile and the curve of sectional areas and other hydrostatic data.

A-Tirma G2 has a length overall (LOA) of two meters, and in opposition to A-Tirma first generation, it is provided with two wing sails. In Fig. 4 is shown a stern view of the boat. Its design has been conceived with a stability curve which is positive for all heel angles, as we can observe in Fig. 5. Furthermore, its smooth slope produces a relatively soft behavior when an increase in wind speed makes her heel. Due to its nearly cylindric hull shape, the curve has a relatively smooth slope, and it does not have a large righting moment for the first 85° of heeling, obtaining a relatively smooth response to heeling efforts due to the action of wind on the sails. This contributes to reduce the stress on rigging.

The keel is a trapezoidal NACA63-009 [3] profile with a retracted bulb with a NACA 0012 [3] section, optimized to avoid entanglement with floating debris and seaweeds.

Table 1 A-Tirma G2's main dimensions

Dimension	Value (m)
Sketch length	2.000
Total length	1.985
Sketch beam	0.370
Maximum beam	0.488 (including skegs and rudders)
Sketch draught	0.628

Table 2 A-Tirma G2's properties

Volume properties		**Initial stability**	
Displaced volume (Vol.)	0.042 m^3	Transverse metacentric height (KMt)	0.637 m
Displacement (Dsp.)	0.043 tonnes	Transverse metacentric height (BMt)	0.075 m
Area of wet surface (Sw)	1.224 m^2	Longitudinal metacentric height (KMl)	2.907 m
Length pos. of hull center (LCB)	1.021 m	Longitudinal metacentric height (BMl)	2.345 m
Length pos. of hull center (LCB)	2.458	**Lateral plan**	
Transverse pos. of hull center (TCB)	0.000 m	Lateral area	0.322 m^2
Vertical pos. of hull center (KCB)	0.562 m	Longitudinal pos. of center of efforts	1.040 m
Middle section properties		Vertical pos. of center of efforts	0.436 m
Middle section area (Sm)	0.077 m^2	**Features of hull above waterline**	
Waterline properties		Lateral wind exposed area	0.720 m^2
Length in waterline (Lwl)	1.946 m	CoG's z coordinate of wind exposed area	1.063 m
Beam in waterline (Bwl)	0.367 m	CoG's x coordinate of wind exposed area	1.050 m
Waterline surface (Swl)	0.543 m^2	CoG's x coordinate of wind exposed area	1.050 m
Length pos. center of waterline (XF)	1.008 m	Distance from bow of wind exposed area CoG	0.896 m
Transverse pos. center of waterline (TF)	0.000 m	**Weight: 0.043 t**	
Entry angle (Beta)	1.434°	**CoG X: 1.021 m CoG Y: 0.000 m CoG Z: 0.469 m**	
Transverse moment of inertia (It)	0.003 m^4		
Longitudinal moment of inertia (Il)	0.099 m^4		

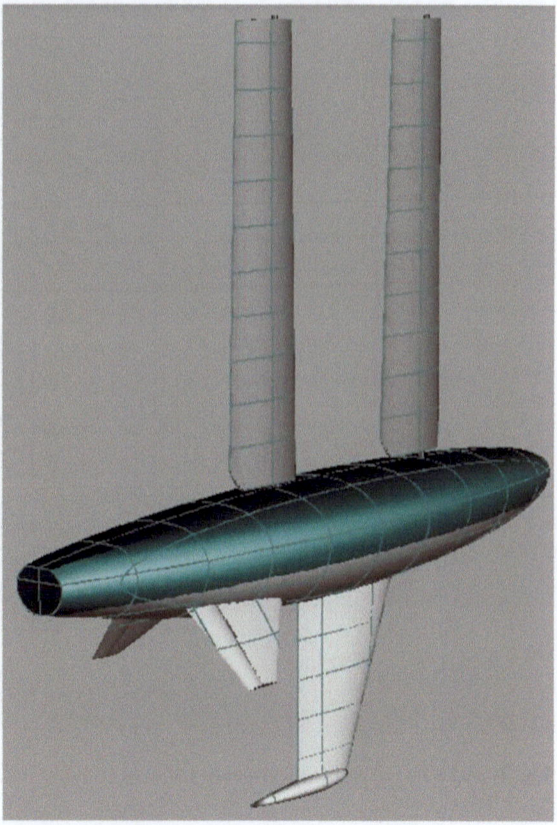

Fig. 2 A-Tirma G2 perspective. Cilindric section, two wing sails and two 40° rudders with skeg

For heeling angles of 90° or more, wing sails contribute to prevent capsizing and facilitate the boat to get to an upright position, in case of capsizing due to a wave. In addition, in order to have a better resistance to capsize and more directional stability, A-Tirma G2 presents a proportional bigger lateral surface resistance than many actual boats due to its skegs and keel, although at the cost of sacrificing some speed due to an increase in drag.

The efforts to improve the directional stability for all wind intensities and heeling angles, allow to optimize the power consumption dedicated to govern the sailboat.

The behavior of the torque produced on a classic sailboat with a single rudder is different from A-Tirma G2. The *LEAD*, longitudinal distance between the center of lateral resistance and the center of efforts on the sails, should be studied for an optimal point of heeling, at which the torque is canceled and the rudder keeps amidships. A-Tirma G2 can not reef their sail to set her heeling. On the contrary, it has two rudders and skegs that move aft the CLR when the boat is heeling. This contributes to cancel the weather helm and improve course stability (Fig. 6).

Each rudder will actuate in a given range of heeling angles at each side, overlapping its actuation at 5° on each range. Both rudders, albeit effective only when

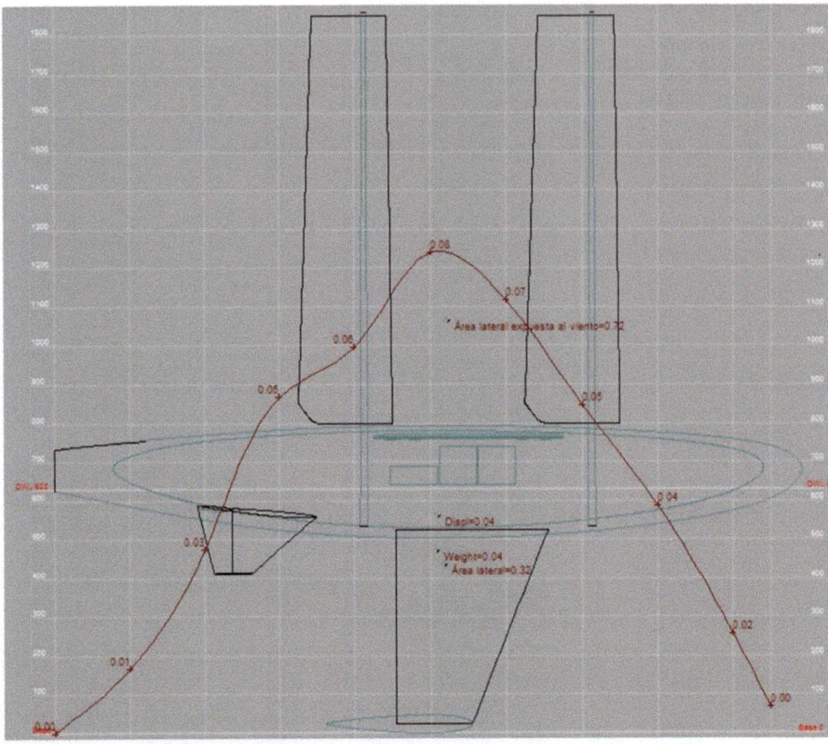

Fig. 3 A-Tirma G2 profile view and its curve of sectional areas and other hydrostatic data

sailing heeled on one side, allow to reach a long term final goal, even under precarious navigation conditions. In Fig. 7 we can observe how only one rudder actuates for high heeling angles.

Wing sails have been built on carbon fiber using a symmetrical NACA 0009 [3] profile with a wingspan of 1.05 m and mean of chord of 0.225 m, equivalent to an aspect ratio of 4.6.

The choice of two semi-balanced or compensated wing sails follows several objectives. Namely, like using two rudders, they allow to keep navigating in case of breakdown of one of them. Moreover, two sails instead of one, with equivalent surface, produces less heeling moment as the sail plan center descends. Thus, the behavior downwind and in strong winds is improved. Also compensation contributes to minimize power consumption, as it reduces the torque needed to trim them.

Benatar et al. [4] have demonstrated that a rudderless sailboat with a double mast configuration can be headed using only the sails. Considering this possibility, a rig of two wing sails may also be interesting as a final resource for steering the boat in case of total failure of the rudders.

Fig. 4 A-Tirma G2.
Stations view

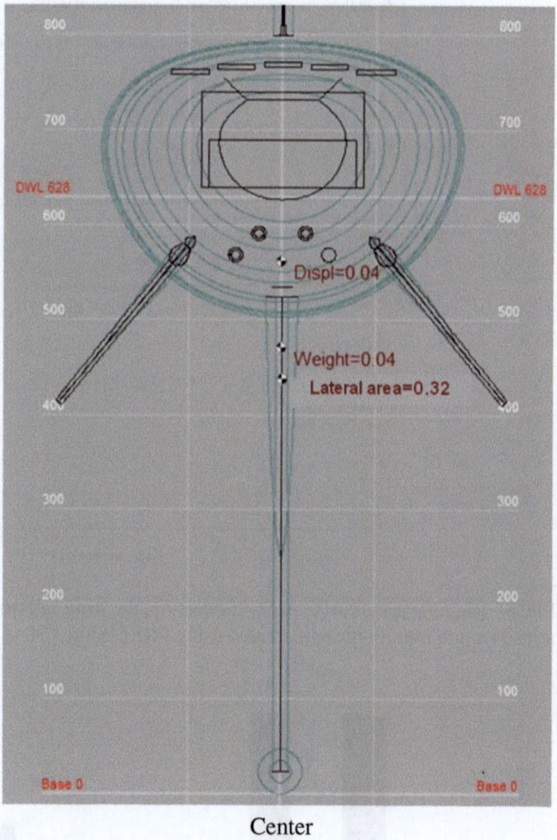

Note also that for navigating downwind, rigid sails are more efficient than cloth sails on any standard rig because at this point of sailing, cloth sails produce thrust entirely by drag, which clearly depends on the magnitude of the apparent wind. On the other hand, when sailing downwind using wing sails, thrust is obtained from lift and, maybe, also from drag. Figure 8 illustrates both situations.

In relation to the drive system for the wing sails, having a system under deck would have been preferable, being free of possible entanglements with floating debris and algae. Nevertheless, in this first operating prototype of A-Tirma G2, we have installed a traditional sheet system for RC (Remote Controlled) sailboats based on a servo with a drum.

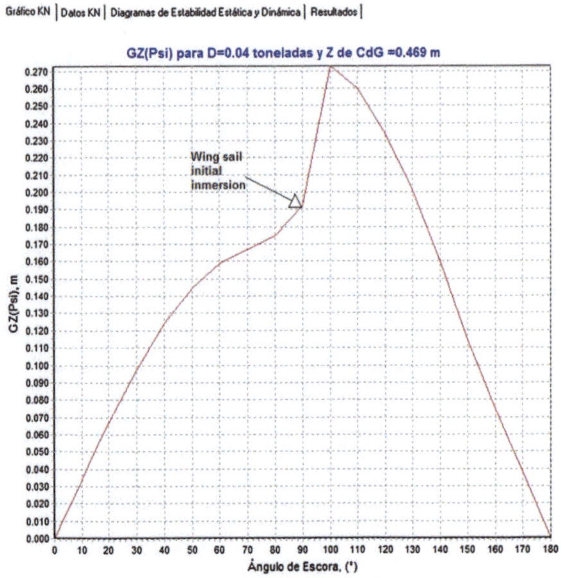

Fig. 5 Righting arms curve. At 90° of heeling, the wing sails touch water and the keel exists. Right arms increase rapidly (design created using FREE!Ship 3.30+)

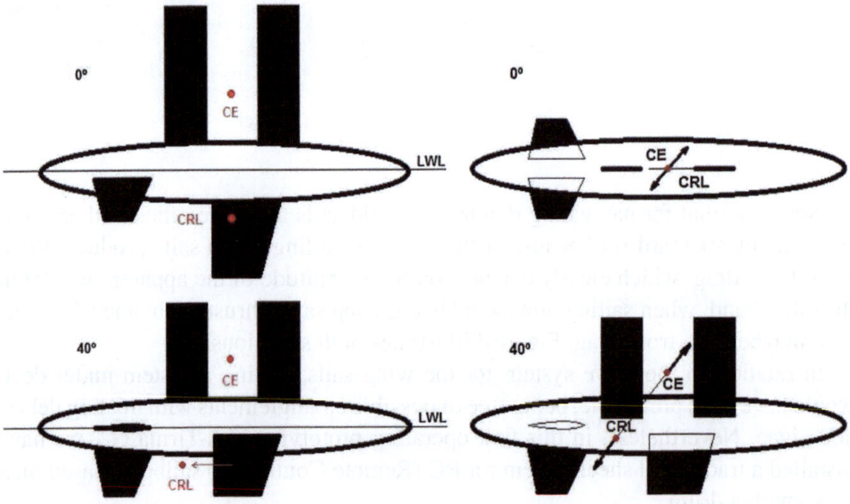

Fig. 6 A-Tirma G2. Sketch of the approximate equilibrium of forces, generated on the sails and the lateral resistance at 0° and 40° degrees of heeling. The configuration of skegs and rudders provides an equilibrium of moments in a wide range of heelling angles. On the figures, CE is the Center of Efforts on the sails, and CRL is the Center of Lateral Resistance

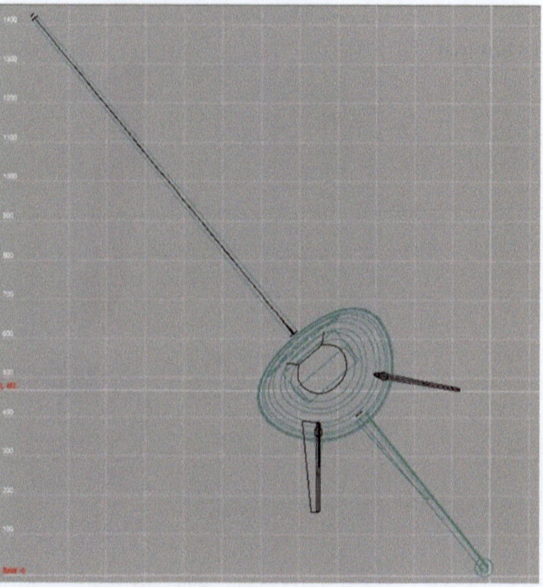

Fig. 7 A-Tirma G2. Only one rudder actuating for a heeling of 40° degrees

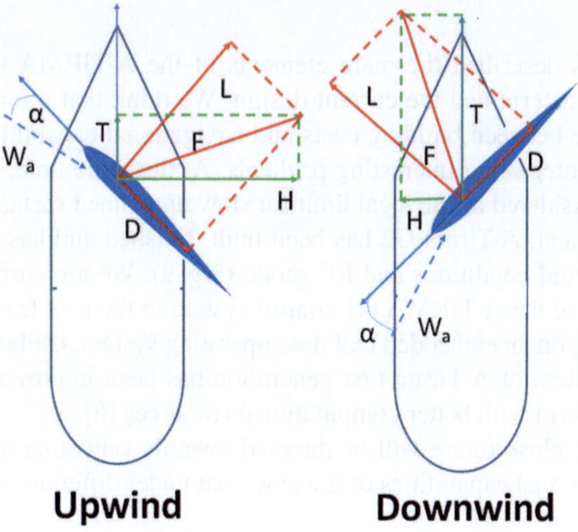

Fig. 8 Wing sail configurations for sailing upwind and downwind, where W_a is the apparent wind, alpha is the angle of attack of the profile with respect to apparent wind, L and D are the lift and drag produced by the wing profile. R is the resultant force that decomposes in H, a heeling component, and T, the thrust that effectively propels the vessel

Fig. 9 A-Tirma G2 prototype during its first trial at sea

3 Summary

This paper has described the main elements of the A-TIRMA G2 and the rationals that have determined the current design. We think that a 2 meter LOA offers a good balance between building costs and navigational capabilities, including the possibility of integrating interesting payloads. At the same time, the 2 meter LOA starts to be considered as the legal limit for slow unmanned surface vehicles at sea.

At the moment, A-Tirma G2 has been built, finished and has passed initial sea trials in low wind conditions and RC mode (Fig. 9). We are currently working on the adaptation of the A-TIRMA G1 control system to the new features of the G2. It will be running on an embedded real time operating system, ChibiOS/RT [5], where the control system of A-Tirma first generation has been improved and ported to a hardware platform with better computational resources [6].

Work in the close future will be directed towards validating the control system and the navigational capabilities of the new boat under different sea conditions.

Acknowledgments The authors are sincerely grateful to Solumatica Canarias for providing financial support for building the A-TIRMA G2 prototype.

References

1. Cabrera-Gámez J, Ramos-de Miguel A, Domínguez-Brito A, Hernández-Sosa J, Isern-González J, Fernández-Perdomo E (2013) An embedded low-power control system for autonomous sailboats. In: Bars FL, Jaulin L (eds) Robotic sailing 2013: poceedings of the 6th international robotic sailing conference. Springer International Publishing, pp. 67–79. doi:10.1007/978-3-319-02276-5_6
2. A-Tirma's Blog (2013) http://velerorobot.blogspot.com.es
3. Goett HJ, Bullivant WK (1939) Tests of NACA 0009, 0012 and 0018 airfoils in the full-scale tunnel. In: NACA technical report 647 (NACA-TR-647). National advisory committee for aeronautics. Langley aeronautical lab. Langley Field, VA, United States. http://ntrs.nasa.gov/search.jsp?R=19930091723
4. Benatar N, Qadir O, Owen J, Baxter P, Neal M (2009) P-controller as an expert system for manoeuvring rudderless sail boats. In: UK workshop on computational intelligence (UKCI 2009). University of Nottingham, UK, 7–9 Sept 2009
5. ChibiOS/RT (2015) Home page www.chibios.org
6. Cabrera-Gámez J, Ramos-de Miguel A, Domínguez-Brito AC, Hernández-Sosa JD, Isern-González J, Adler L (2014) A real-time sailboat controller based on chibios. In: Morgan F, Tynan D (eds) Robotic Sailing 2014: Proceedings of the 7th international robotic sailing conference. Springer International Publishing, pp. 77–85. doi:10.1007/978-3-319-10076-0_7

Integration of Wind Propulsion in an Electric ASV

Nuno A. Cruz, José C. Alves, Tiago Guedes, Rômulo Rodrigues, Vitor Pinto, Daniel Campos and Duarte Silva

Abstract This paper describes the steps taken to integrate wind propulsion in the Zarco ASV, a small size electric powered catamaran. In terms of hardware, the original structure of the vehicle has been enlarged to accommodate one or more sails, and the proper interfaces have been included to allow measurement of wind speed and direction and independent sail actuation. The sails are lightweight rigid wings assembled from a core of balsa wood, reinforced with aluminum and epoxy. Each sail angle can be controlled by a servo, commanded by the main CPU and taking into account the wind speed and direction, as well as the lift and drag curves of the wing sails, according to some predefined control strategy.

1 Introduction

Robotic surface vehicles are extremely valuable scientific devices that have been playing a consistent role in many areas of marine science, by providing an effective and affordable way to sample the ocean. These vehicles are commonly referred to as

N.A. Cruz (✉) · J.C. Alves (✉)
FEUP/INESC TEC, Rua Dr Roberto Frias, 4200-465 Porto, Portugal
e-mail: nacruz@fe.up.pt

J.C. Alves
e-mail: jca@fe.up.pt

T. Guedes · R. Rodrigues · V. Pinto · D. Campos · D. Silva
FEUP, Rua Dr Roberto Frias, 4200-465 Porto, Portugal
e-mail: ee12239@fe.up.pt

R. Rodrigues
e-mail: ee09264@fe.up.pt

V. Pinto
e-mail: ee08262@fe.up.pt

D. Campos
e-mail: ee08182@fe.up.pt

D. Silva
e-mail: ee08212@fe.up.pt

A. Friebe and F. Haug (eds.), *Robotic Sailing 2015*,
DOI 10.1007/978-3-319-23335-2_2

Fig. 1 Zarco typical configuration, with two independent electrical thrusters in the stern

Autonomous Surface Vehicles (or *ASVs*), and they combine the ability to transport a large variety of sensors and actuators with real time, high-bandwidth communications, resulting in advanced systems for tele-presence in the marine environment.

There are several options to provide propulsion to an ASV, both in terms of technology involved and also in terms of configuration, and there are several successful examples using these different options. Zarco is a small size electric powered Autonomous Surface Vehicle, in operation at the University of Porto since 2005 [1]. It can be seen in Fig. 1 in its original configuration, with an overall length of 1.5 m, weighting approximately 50 kg and propelled by two independent electrical thrusters, with a maximum velocity of 4 knots. Our default choice was a configuration with differential drive provided by modified COTS trolling motors, due to a combination of price, availability and simplicity of control. These thruster are relatively inexpensive, available of the shelf from several manufacturers in a variety of power levels, and there are also many off-the-shelf solutions of electronic boards to control them from a computer.

In a typical operation, the electric propulsion of Zarco accounts for about 90 % of the power consumption, yielding about 4–6 h of autonomy at a nominal velocity of 2 knots. This percentage depends on many operational factors, like velocity and payload, but it is largely dominated by the hydrodynamic drag, therefore it is always a significant portion of the total power of any ASV mission. In the case of Zarco, the vehicle construction is highly modular and the energy enclosure can be swapped for a fresh pack to extend the mission duration. However, there are many situations in which the vehicle has to operate without interruptions, and therefore the possibility of integrating wind propulsion and harvesting energy from the environment to increase autonomy are very appealing as they open up the possibility of permanent ocean presence.

The remaining of the paper is organized as follows. In Sect. 2, we provide some background on technologies of wind propulsion, mainly intended for robotic vehicles. Then, in Sect. 3, we detail the construction of a wing sail and report its characteristics as an independent propulsion system. We proceed, in Sect. 4, to the steps taken to integrate such a propulsion system in Zarco, and, in Sect. 5, we discuss some possibilities of taking advantage of the hybrid propulsion system. Finally, we draw some conclusions regarding the accomplished work and we provide some ideas regarding the tasks planned for the near future.

2 Background and Related Work

2.1 Using Wind for Propulsion

The wind has been the major source of ship propulsion for thousands of years, until the advent of fossil fuels. Recently, with the rising prices of oil and the urgent need to reduce CO_2 production, the use of wind for ship propulsion has been regaining a growing attention. Some new concepts have been proposed for harvesting wind energy [8] and some companies are already delivering products for new or existing vessels. For example, SkySails GmbH, in Germany, offers wind propulsion systems as an auxiliary fuel-saver for cargo ships, based on a large towing kite [19], while Eco Marine Power Co. Ltd., in Japan, is developing the EnergySail, a rigid sail that can be fitted with solar panels to take advantage of both wind and solar energy [3].

In the case of robotic systems, the typical scale is much smaller, but the prospect of using wind as a major source of energy for ASV propulsion is also very appealing as it reduces a very significant portion of the total energy delivered by the onboard batteries during a mission. Moreover, the remaining electronics typically require such a small amount of energy that it can be provided by a simple management system based on rechargeable batteries fed by a renewable source (for example, a solar panel). However, the use of this infinite source of energy for the propulsion of a robotic system still presents a multi-disciplinary challenge that is being addressed in the latest years. Most wind propelled ASVs employ either a fabric-based or a wing sail, each with its own pros and cons, both from the performance point of view, but also in practical terms [13]. Fabric based sails, or *soft* sails, mimic the usual sails of standard sailboats, but are very difficult to monitor in terms of 3D shape and automatic detection of improper tuning, like luffing. Although there has been some work to detect the correct trimming of soft sails [12], it is easier to control single-element, symmetrical wing sails from a robot, with very reasonable performance [4].

2.2 Wind Propelled ASVs

Autonomous Sailboats are a particular class of ASVs that rely on wind to provide propulsion and only need electrical energy for the onboard electronics and rudder/sail adjustments. They have gained particular attention in the last few years for their unique ability to maintain long term unassisted operations in the sea surface, with a wide range of application scenarios spanning the scientific, civil, or military communities [2, 15].

One important point to address for the permanent ocean presence is the ability to withstand the harsh conditions at sea during long periods of time, and a few recent successful projects have demonstrated the capability of long range navigation with robotic autonomous sailboats. The Saildrone project is one example, a 19 ft long trimaran rigging a wing sail, demonstrated in the end of 2013 with a 100 day autonomous mission in the Pacific Ocean, and an announced plan to try a circumnavigation [16]. Other projects of small sea worth autonomous sailing robots have also been developed in Europe, as the Austrian ASV Roboat exploited in research on marine mammals [20], the BeagleB project of the University of Aberystwyth, United Kingdom [17], the Vaimos sailboat from ENSTA Bretagne, France [10], and the Portuguese FASt, used for acoustic monitoring of mammal activity [18]. In terms of hybrid propulsion for small ASVs, the University of Aberystwyth has developed a 2 m long prototype of a hybrid sail and electric drive boat [11], but unfortunately it has not evolved into the intended full-scale oceanographic version.

On a related application scenario, the utilization of wind as a complementary source of propulsion has also been proposed to ensure station keeping of buoys, as in the case of the Norwegian SailBuoy vessel, from CMR Instrumentation [5], or the *Station Keeping Buoy* from John Hopkins University, in the USA [6].

On a more commercial front, Harbor Wing Technologies, in the USA, developed a few prototypes of fully wind propelled multi-hull surface vehicles, mainly intended for the US Navy, but the company stopped reporting any new developments in 2010 [7]. Some of their key personnel later founded Ocean Aero Inc., also in the USA, a company that has announced the development of the *Submaran*, a 4.1 m long hybrid surface/sub-surface vessel, powered by wind and solar energy [14].

3 Design and Construction of the ASV Sail

3.1 Design Considerations

Given the pros and cons of fabric-based sails and wing sails, we've decided to take advantage of the simplicity of implementation and repeatability of wing sails behavior. Our initial requirements for the sail characteristics were mainly dictated by physical constraints, not only in terms of manufacturing and mechanical integration, but also in terms of logistics. Ideally, a large sail area provides the most actuation and this may be achieved by a long wing sail with a large cord. On the other hand, a longer sail will increase wind-induced pitch and roll and, at the same time, will shade a similar sail in the case of using multiple instances.

We've decided to limit the length of the sail to one meter and the cord to 40 cm, yielding a design that we could easily transport, if detachable from the main structure. We've also analyzed the impact of the aspect ratio in construction and control and we've decided to use a NACA 0015 profile, a well studied design with a symmetrical shape enabling similar lift generation in both sides of relative orientation.

3.2 *Sail Construction*

The first sail prototype was assembled from a core of balsa wood with aluminum and epoxy reinforcements and wrapped in a thermal foil. Figure 2 shows some of the construction stages. To ensure the overall NACA 0015 profile, 10 cross sections were machined from 5 mm thick balsa wood, to become the backbone of the sail. These were drilled with guiding holes to ensure a proper alignment: a central hole with 20 mm diameter for an aluminum tube that is used as the sail mast, and smaller holes for 8 mm aluminium tubes that reinforce the structure. To facilitate construction and avoid sharp edges, we've modified the leading and trailing edges of the sail to round shapes. The leading edge is then provided by a 16 mm PVC tube and the trailing edge is formed by a 4 mm carbon rod.

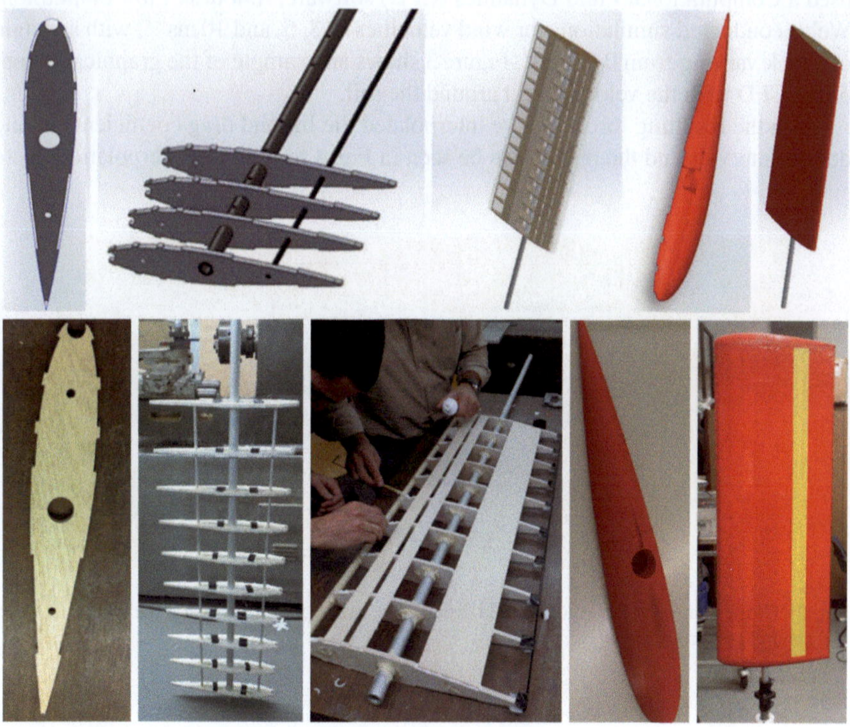

Fig. 2 Different stages of the construction of the wing sail prototype, from the CAD drawings to the final result

The wooden cross sections were held in place using aluminum rods evenly spaced along the sail mast, and passing through it, and everything was held together with epoxy, resulting in a very sturdy backbone. On the outside, this backbone had 8 slots (4 on each side) meant to hold balsa wood battens, in order to provide an outer shell with a profile close to NACA 0015. These battens were also glued with epoxy before being wrapped with the resin-coated thermal foil, which was carefully glued to the wood using a heatgun.

With this assembly, both the top and bottom of the wing sail had sharp edges, which could cause turbulence and affect aerodynamics. To avoid this, we've designed two mechanical parts to round both ends of the sail, and we've used a 3D printer to fabricate them in ABS plastic. The final assembly weights 1.3 kg and it can be seen in the last picture of Fig. 2.

3.3 Aerodynamic Characteristics of the Sail

In order to evaluate the aerodynamic characteristics of a single sail under various wind profiles and compare it to the theoretical values of a NACA 0015 profile, we've used a Computational Fluid Dynamics (CFD) software, *Autodesk Flow Simulation*. We've conducted simulations for wind velocities of 3, 5, and 10 ms^{-1}, with an angle of attack varying from 0 to 180°. Figure 3 shows an example of the graphical output of the CFD with the velocity field around the sail.

With the resulting forces, we've interpolated the lift and drag coefficients in one degree intervals and the results can be seen in Fig. 4 for a linear interpolation. Note

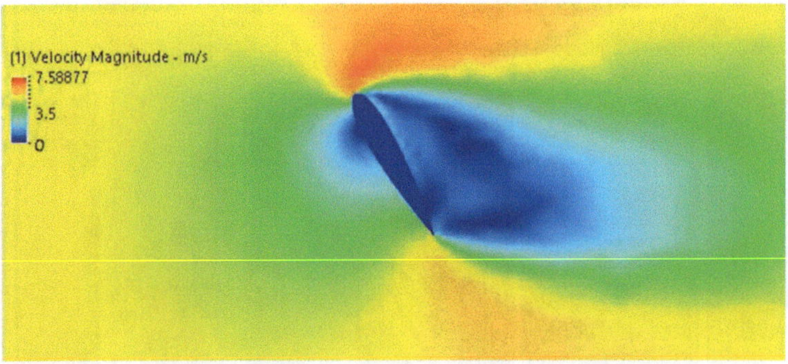

Fig. 3 Cross section of the velocity field around the wing sail, simulating a 5 ms^{-1} wind speed and an angle of attack of 60°

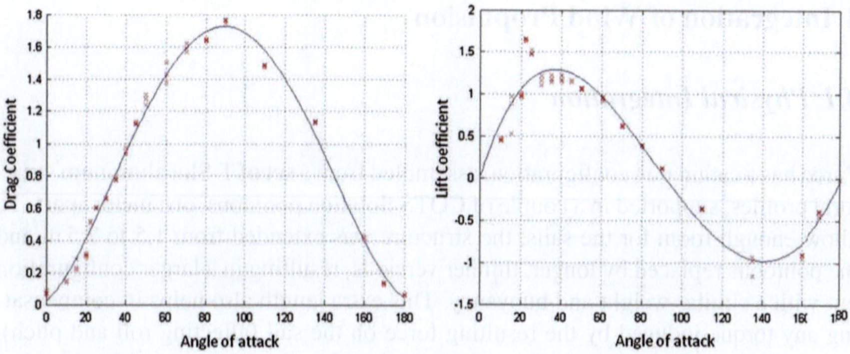

Fig. 4 Interpolation of drag and lift coefficients obtained by CFD analysis of the wing sail

that this linear interpolation of the lift coefficients has a significant error around 20–30°, therefore a practical implementation has to use a higher order interpolation to yield more accurate results.

With these coefficients, it is possible to estimate the drag and lift forces provided by a sail, for each apparent wind angle and magnitude. The combined effect of these forces represent both the magnitude and the angle of the actuation force provided by the sail. As an example, the plot of Fig. 5 shows the magnitude (in Newton) and the angle of the resulting force, when the wing sail is under a 10 ms^{-1} wind.

Fig. 5 Magnitude and angle of the combined drag and lift forces, for different angles of attack

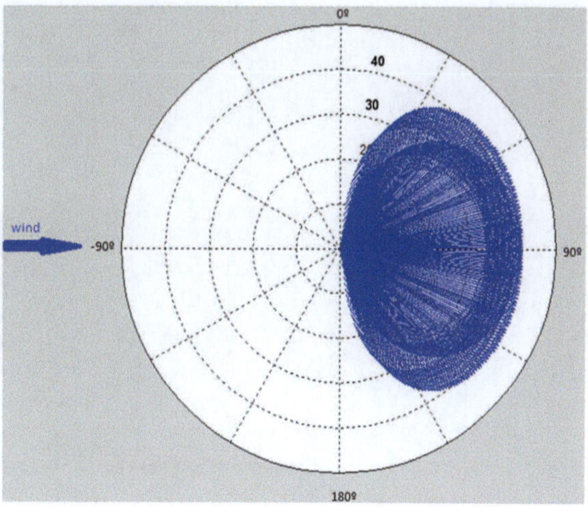

4 Integration of Wind Propulsion

4.1 Physical Integration

Zarco has a catamaran configuration, assembled from a set of T-Slot aluminum structural profiles, supported in a couple of COTS flotation pontoons, one meter apart. To allow enough room for the sails, the structure was extended from 1.5 to 2.5 m and the pontoons replaced by longer, thinner versions, resulting in a larger configuration but with a similar weight and buoyancy. This extra length also helps in compensating any torque induced by the resulting force on the sail (affecting roll and pitch). Each sail angle can be controlled by a waterproof servo, commanded by the main CPU using a simple servo motor controller. In order to measure the wind speed and direction, we've installed an Airmar 150WX weather station, that outputs NMEA sentences into a CPU serial port. Figure 6 shows the configuration of Zarco with a wing sail at the stern, during initial testing.

4.2 Hybrid ASV Model

The integration of additional propulsion to the Zarco ASV results in a small change in the model of the vehicle. Note that we assume that Zarco is equipped with an

Fig. 6 Zarco being tested with a single wing sail at the stern

electrical propulsion system that is able to answer adequately to its dynamic and kinematic tasks, described in [9] for the particular case of positioning control. The installation of sails aims at lowering the power consumption, consequently enlarging the endurance of the vehicle. The number of sails is restricted by physical limitations of the boat and disturbance that may arise if two or more are placed close together.

Take a body fixed coordinate system $\{B\}$ fixed to the center of mass of the vehicle, whose longitudinal axis indicates the surge speed and transversal axis the sway speed. The thrusters, placed on the rear of the vehicle, are responsible for the force on the longitudinal axis $X_{act\ motor}$ and torque $N_{act\ motor}$. Each i_{th} sail produces forces in the longitudinal $X^i_{act\ sail}$ and transversal axis $Y^i_{act\ sail}$ and a torque $N^i_{act\ sail}$. Consequently the input actuation can be described as:

$$X_{act} = X_{act\ motor} + \sum_i^n X^i_{act\ sail} \tag{1}$$

$$Y_{act} = \sum_i^n Y^i_{act\ sail} \tag{2}$$

$$N_{act} = N_{act\ motor} + \sum_i^n N^i_{act\ sail} \tag{3}$$

Note that the sails allow actuation in the transverse axis (the term Y_{act}) which was not possible when propulsion was provided only by thrusters. However, the sail actuation in this transverse axis may quickly saturate and the catamaran hull configuration offers much resistance to lateral motion. Therefore, no reference should be set to Y_{act} as, from the control perspective, the vehicle remains underactuated. Note also that the forces produced by the sails result in pitch and roll moments that cause the vehicle to heel, but, fortunately, our catamaran configuration minimizes such effects and therefore they can be neglected in our equations.

Consider the wind coordinate system $\{W\}$ fixed to the aerodynamic center of a sail, where x_w is aligned with the apparent wind direction. Let θ^B_W denote the angle between the vessel and the wind, and $R(\theta^B_W)$, the rotation matrix from frame $\{W\}$ to $\{B\}$. The actuation provided by each sail is then

$$\begin{bmatrix} X^i_{act\ sail} \\ Y^i_{act\ sail} \end{bmatrix}_B = R(\theta^B_W) \begin{bmatrix} F^i_D \\ F^i_L \end{bmatrix}_W \tag{4}$$

$$N^i_{act\ sail} = \begin{bmatrix} X^i_{act\ sail} & Y^i_{act\ sail} \end{bmatrix}_B \begin{bmatrix} pos^i_x \\ pos^i_y \end{bmatrix} \tag{5}$$

where F^i_D and F^i_L are the drag and lift forces, respectively, and (pos^i_x, pos^i_y) are the position of the sail wrt the center of mass.

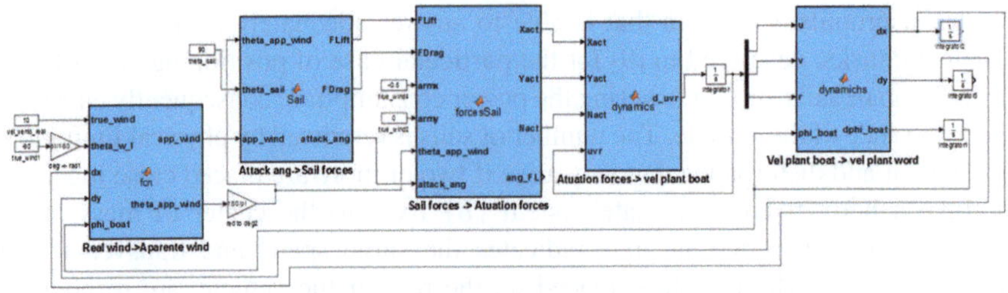

Fig. 7 Complete simulation model of the hybrid ASV developed in Simulink

Note that in the original electrical system, the actuation was simply given by:

$$X_{act\ motor} = \begin{bmatrix} F_{port} & F_{starboard} \end{bmatrix} \begin{bmatrix} 1 \\ 1 \end{bmatrix} \tag{6}$$

$$N_{act\ motor} = \begin{bmatrix} F_{port} & F_{starboard} \end{bmatrix} \begin{bmatrix} \frac{dist_{motor}}{2} \\ -\frac{dist_{motor}}{2} \end{bmatrix} \tag{7}$$

In order to simulate the dynamics of the ASV under different control strategies, we've developed a model of the various blocks in Simulink, shown in Fig. 7.

5 Strategies for Propulsion of a Hybrid ASV

The wing sails can be considered as additional inputs for the propulsion system, as described in the previous section, but with the remark that its characteristics are dynamic, depending on apparent wind speed. In this section, we provide two examples of possible uses of the wing sails in different application scenarios.

5.1 Using the Wind as the Main Propeller

Given that the Zarco ASV does not have any rudder, the use of a single sail can only be effective for apparent wind angles up to 90 degrees. In many situations, however, this may be sufficient to bring the vehicle to a safe area if the batteries are at a critical level, or in case of some other major failure.

In this case, the strategy to use is to compute the required direction of motion and then search for the closest direction of the combined drag and lift forces, according to the real time measurement of the wind direction.

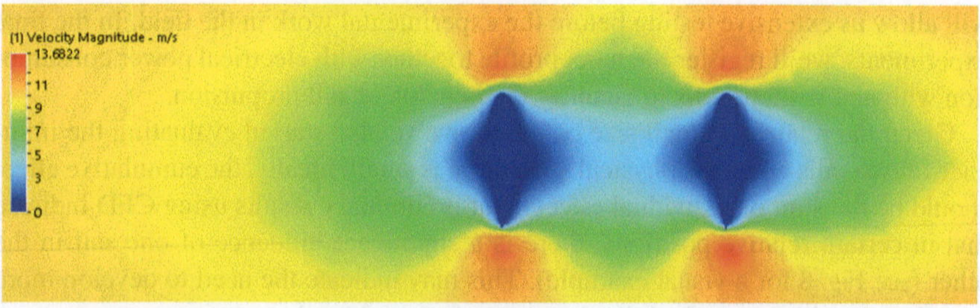

Fig. 8 Cross section of the velocity field around two wing sails separated by 2 cord-lengths (80 cm), simulating a $10\,\mathrm{ms}^{-1}$ wind speed and an angle of attack of 90°

5.2 *Coupled Wind and Electrical Thrust*

One of the most promising uses of the sail(s) is to provide extra propulsion to alleviate the power required by the electrical thrusters. In practical terms, the problem can be formulated as determining the angle of attack of the sail that given an actuation reference $\tau_{ref} = [X_{act}, N_{act}]$, minimizes the power required by the thrusters.

The nonlinear model of the drag and lift forces increase the complexity of a classic control approach. Moreover, it may happen that the minimization of total power required by the electrical is obtained when the resulting forces of the sails are not pointing towards a given target. Therefore, the method proposed relies on a searching algorithm, considering that the sail can be oriented in a finite number of angles of attack. For each of these angles, we can calculate the drag and lift forces, and the sail actuation on the body fixed frame. We can then compute the complementary propulsion that the thrusters have to deliver in order to follow the specified reference. Finally, we choose the sail orientation that corresponds to the minimum power consumption.

6 Conclusions and Future Work

This paper describes the steps taken to integrate wind propulsion in Zarco, evolving from a purelly electric ASV into a hybrid vehicle. We've built a wing sail prototype with one meter of height and 40 cm of cord. A CFD software was used to simulate this sail in various wind conditions, to extract the hydrodynamic coefficients and therefore characterize the sail propulsion as a function of the wind. This sail has been physically installed in Zarco and it is currently being tested in the field. After validation of the expected performance, a second wing sail will be built to provide further thrust and more directional control.

The paper also provides some ideas on how to explore the available propulsion, and these will be explored in the near future. The validation of these techniques will benefit from the full dynamic model of the vehicle, available in Simulink, which will allow us extensive testing before the experimental work in the field. In the final experiments, we'll register the wind profile together with electrical power consumption with and without sails to quantify the benefits of sail propulsion.

Given the relatively small size of Zarco, we've also started evaluating the influence of two sails when the physical separation is small. Ideally, the cumulative effect should be the sum of individual effects, but preliminary results using CFD indicate that in certain relative positions, there is a significant influence of one sail in the other (see Fig. 8 for a visual example). This may indicate the need to develop more complex algorithms to take full advantage of multiple sails.

Finally, we will also analyze the influence of the wing sails in terms of pitch and roll moments. Although the catamaran configuration of Zarco minimizes the impact of such moments, this analysis may be useful for other hull configurations.

Acknowledgments This work is financed by the ERDF—European Regional Development Fund through the COMPETE Programme (operational programme for competitiveness) and by National Funds through the FCT—Fundação para a Ciência e a Tecnologia (Portuguese Foundation for Science and Technology) within project "FCOMP-01-0124-FEDER-037281". The authors are also thankful to the anonymous reviewers who provided valuable suggestions to improve the final version of the manuscript.

References

1. Cruz N, Matos A, Cunha S, Silva S (2007) Zarco—an autonomous craft for underwater surveys. In: Proceedings of the 7th geomatic week. Barcelona, Spain
2. Cruz NA, Alves JC (2008) Autonomous sailboats: an emerging technology for ocean sampling and surveillance. In: Proceedings of the MTS/IEEE international conference on oceans'08. Quebec, Canada
3. Eco Marine Power Co. Ltd.: EnergySail. http://www.ecomarinepower.com/en/energysail (accessed May 2015)
4. Elkaim GH, Boyce Jr CO (2007) Experimental aerodynamic performance of a self-trimming wing-sail for autonomous surface vehicles. In: Proceedings of the IFAC conference on control applications in marine systems, IFAC CAMS 2007. Bol, Croatia
5. Fer I, Peddie D (2013) Near surface oceanographic measurements using the SailBuoy. In: Proceedings of the MTS/IEEE international conference on oceans'13. Bergen, Norway
6. Frizzell-Makowski LJ, Shelsby RA, Mann J, Scheidt D (2011) An autonomous energy harvesting station-keeping vehicle for persistent ocean surveillance. In: Proceedings of the MTS/IEEE international conference on oceans'11, pp 1–6. Kona, HI, USA
7. Harbor Wing Technologies, Inc.: http://www.harborwingtech.com/ (accessed May 2015)
8. Kantharia R (2013) Top 7 green ship concepts using wind energy. Marine Insight. http://www.marineinsight.com/marine/marine-news/headline/top-7-green-ship-concepts-using-wind-energy/
9. Matos A, Cruz N (2008) Positioning control of an underactuated surface vessel. In: Proceedings of the MTS/IEEE international conference on oceans'08. Quebec, Canada
10. Ménage O, Bethencourt A, Rousseaux P, Prigent S (2013) Vaimos: realization of an autonomous robotic sailboat. In: Robotic sailing 2013. Springer, pp 25–36
11. Miller P, Sauz C, Neal M (2014) Development of ARRTOO: a long-endurance, hybrid-powered, oceanographic research vessel. In: Bars FL, Jaulin L (eds) Robotic sailing 2013. Springer International Publishing, pp 53–65

12. Murray-Davis H, Barrett D (2015) Piezoelectric vibrational sensor for sail luffing detetection on robotic sailboats. In: Morgan F, Tynan D (eds) Robotic sailing 2014. Springer International Publishing, pp 87–93
13. Neal M, Sauz C, Thomas B, Alves JC (2009) Technologies for autonomous sailing: wings and wind sensors. In: Proceedings of the 2nd international robotic sailing conference. Matosinhos, Portugal
14. Ocean Aero Inc.: http://www.oceanaero.us/documents/OAHandout15MetricFeb172015.pdf (accessed May 2015)
15. Rynne PF, von Ellenrieder KD (2009) Unmanned autonomous sailing: current status and future role in sustained ocean observations. Mar Tech Soc J 43(1):21–30
16. Saildrone Inc.: http://www.saildrone.com (accessed May 2015)
17. Sauzé C, Neal M (2008) Design considerations for sailing robots performing long term autonomous oceanography. In: International robotic sailing conference, Breitenbaum, Austria, pp 21–29
18. Silva A, Matos A, Soares C, Alves J, Valente J, Zabel F, Cabral H, Abreu N, Cruz N, Almeida R, Ferreira RN, Ijaz S, Lobo V (2013) Measuring underwater noise with high endurance surface and underwater autonomous vehicles. In: Proceedings of the MTS/IEEE international conference on oceans'13. San Diego, CA, USA
19. SkySails GmbH: http://www.skysails.info/english/ (accessed May 2015)
20. Stelzer R, Jafarmadar K (2012) The robotic sailing boat ASV Roboat as a maritime research platform. In: Proceedings of 22nd international HISWA symposium

Mobile Autonomous Platforms for Passive-Acoustic Monitoring of High-frequency Cetaceans

Holger Klinck, Selene Fregosi, Haru Matsumoto, Alex Turpin, David K. Mellinger, Anatoli Erofeev, John A. Barth, R. Kipp Shearman, Karim Jafarmadar and Roland Stelzer

Abstract Increased human activities in coastal and offshore waters, including renewable energy efforts such as the deployment and operation of wind, wave, and tidal energy converters, leads to potential negative impacts on marine ecosystems. Efficient monitoring of marine mammals in these areas using stationary passive-acoustic technologies is challenging. Many recreational and commercial activities (e.g., fishing) can hinder long-term operation of moored listening devices. Further, these waters are often utilized by cetaceans such as porpoise species which produce high-frequency echolocation clicks (peak frequency ~130 kHz) for navigation, communication, and prey detection. Because these ultrasonic signals are

H. Klinck (✉)
Cornell Lab of Ornithology, Bioacoustics Research Program, Cornell University,
159 Sapsucker Woods Road, Ithaca, NY 14850, USA
e-mail: Holger.Klinck@cornell.edu

H. Klinck · S. Fregosi · H. Matsumoto · A. Turpin · D.K. Mellinger
NOAA Pacific Marine Environmental Laboratory, Cooperative Institute for Marine Resources Studies, Oregon State University, 2030 SE Marine Science Drive, Newport OR 97365, USA
e-mail: Selene.Fregosi@noaa.gov

H. Matsumoto
e-mail: Haru.Matsumoto@noaa.gov

A. Turpin
e-mail: Alex.Turpin@noaa.gov

D.K. Mellinger
e-mail: David.Mellinger@oregonstate.edu

A. Erofeev · J.A. Barth · R.K. Shearman
College of Earth, Ocean and Atmospheric Sciences, Oregon State University,
104 CEOAS Administration Building, Corvallis, OR 97331, USA
e-mail: aerofeev@coas.oregonstate.edu

J.A. Barth
e-mail: barth@coas.oregonstate.edu

R.K. Shearman
e-mail: shearman@coas.oregonstate.edu

A. Friebe and F. Haug (eds.), *Robotic Sailing 2015*,
DOI 10.1007/978-3-319-23335-2_3

strongly absorbed during propagation, the acoustic detection range is limited to a few 100 m, and therefore the spatial coverage of stationary recorders is relatively limited. In contrast, mobile passive-acoustic platforms could potentially be used to survey areas of concern for high-frequency cetacean vocalizations and provide increased temporal coverage and spatial resolution. In a pilot study, a commercially available acoustic recorder featuring sampling rates of up to 384 kHz was customized and implemented on an autonomous underwater vehicle (AUV) and an unmanned surface vehicle (USV) and tested in the field. Preliminary results indicate that these systems (a) are effective at detecting the acoustic presence of high-frequency cetaceans such as porpoises, and (b) could be a valuable tool to monitor potential negative impacts of renewable energy and other anthropogenic disturbances in the marine environment.

1 Introduction

Increased development and use of marine renewable energy converters harvesting wind, tidal, and wave energy to generate electricity has raised concerns about potential negative impacts of the installation and operation of such devices on the marine environment [4, 10].

In Europe, the offshore wind energy industry is well established. The first commercial windfarm was installed in Vindeby, Denmark in 1991 and many have followed since. To date approximately 2,500 wind turbines are being used in European waters to generate electricity [5]. Wave and tidal energy have been proposed as possible sources of renewable energy in these and other parts of the world. Prototype devices are currently being developed and tested, for example, in the Pacific Northwest of the United States of America [12].

Potential environmental impacts of renewable energy installations are manifold and include the emission of underwater noise [1, 4, 10, 17]. Elevated underwater noise levels are of concern, especially for noise-sensitive cetaceans including the harbor porpoise, *Phocoena phocoena* [16]. Harbor porpoises can be found in temperate and sub-polar coastal waters including the Baltic Sea and the North Pacific [11], and their habitat overlaps with areas of existing and future renewable energy installations. In the North Pacific, the habitat of three additional high-frequency cetacean species, the Dall's porpoise, *Phocoenoides dalli*, and the dwarf and pygmy sperm whale, *Kogia sima* and *K. breviceps* respectively [11],

K. Jafarmadar · R. Stelzer
Austrian Society for Innovative Computer Sciences,
Haussteinstrasse 4/2, A-1020 Vienna Vienna, Austria
e-mail: karim.jafarmadar@innoc.at

R. Stelzer
e-mail: roland.stelzer@innoc.at

overlaps with areas which recently have been proposed for the installation and operation of offshore floating wind turbines.

Monitoring these species is difficult. They are among the smallest cetaceans (body length <3.5 m) and usually (except for the Dall's porpoise) occur in small groups of a few individuals [11, 19] which are hard to spot visually in most weather conditions (Fig. 1c). Conversely, porpoises and *Kogia spp.* regularly emit echolocation clicks [9, 18] for communication, prey detection, and navigation, and these clicks can be readily detected with passive-acoustic monitoring (PAM) systems, regardless of weather or light conditions [9]. However, these ultrasonic signals (peak frequency ~130 kHz; [9, 18] are highly attenuated when propagating due to absorption; therefore, the acoustic detection range is limited to a few 100 m. This limits the effectiveness of stationary acoustic recorders.

Various autonomous underwater vehicles (AUVs) and unmanned surface vehicles (USVs) have been proposed for use in passive-acoustic monitoring efforts [7]. Over the last couple of years AUVs featuring passive-acoustic recording and detection capabilities [2, 6] have proven to be effective survey tools for low- and mid-frequency marine mammal vocalizations. The goal of this study was to evaluate the potential use both AUVs and USVs to monitor high-frequency cetacean

Fig. 1 The USV Roboat **a** and the AUV Seaglider™ **b** The goal of the study was to acoustically detect high-frequency cetaceans including the harbor porpoise **c** Picture credits: **a** Austrian society for innovative computer sciences, Austria, **b** Alfred Wegener Institute, Germany, and **c** Jean-Pierre Bonin, Canada

vocalizations, such as produced by porpoises. These instruments could significantly improve the temporal coverage and spatial resolution of future passive-acoustic survey efforts.

2 Methods

The two vehicles used in this study (Fig. 1a, b) are the Roboat, a prototype autonomous sailboat [15] developed by the Austrian Society for Innovative Computer Sciences (INNOC), Austria, and the Seaglider™ an autonomous deep-diving underwater glider [13], commercially available from Kongsberg, Inc., USA.

Both vehicles were equipped with a commercially available acoustic recorder (Song Meter SM2BAT+, Wildlife Acoustics Inc., USA). The recorders were installed in the science bay of the vehicles, equipped with 896 GB of data storage each (SD memory cards) and programmed to continuously record signals at a sampling rate of 384 kHz and 16 bit resolution. Lossless compression (WAC0) of the audio data was enabled to maximize the available recording duration. Both systems featured a 1 kHz high pass filter and featured a fairly flat frequency response (±10 dB) in the frequency range 1–192 kHz. The Roboat was equipped with a single-ended HTI 96-MIN hydrophone (High Tech Inc., USA) which was mounted to the keel of the boat approximately 0.5 m below the waterline. The overall sensitivity of the acoustic system was −129 dB re 1 V/μPa. The Seaglider was equipped with a differential HTI 92-WB hydrophone (High Tech Inc., USA) mounted to the Seaglider's antenna and a custom-built differential pre-amplifier. The overall system sensitivity was −123 dB re 1 V/μPa. In addition, the acoustic recording system installed on the Seaglider was interfaced with a Persistor CF2 microcontroller (Persistor Instruments Inc., USA) to enable remote control from a base station on shore. The Seaglider also collected environmental data on conductivity, temperature, oxygen, and chlorophyll throughout the mission.

The Roboat test was conducted off of Eckernfoerde, Germany in the Baltic Sea in July 2012. This was the first reported attempt to use an autonomous sailboat to record marine mammal vocalizations. The Seaglider test was conducted off of Newport, OR, USA in the North Pacific in March and April 2014, and was the first reported attempt to sample such high frequencies using a Seaglider.

After recovery of the instruments, collected acoustic data were manually analyzed in the lab. Long-term spectral average plots (LTSAs) were screened for the presence of porpoise echolocation clicks using the Triton software package developed by the Scripps Whale Acoustics Lab, USA (available online at: http://cetus.ucsd.edu/technologies_Software.html).

3 Results

Roboat: The intended transect line for the Roboat was a north-south roundtrip (~130 nm) between Eckernfoerde, Germany and Assens, Denmark. The sea trial was started on 14 July 2012. Unfortunately, the weather conditions during the field test were poor. Average wind speed was measured at 15 kn with gusts of up to 29.5 kn. The Roboat sailed at an average speed of 2.9 kn. After 71 nm of autonomous sailing, severe weather conditions caused a malfunction of the motor necessary to trim the mainsail. Consequently, the trial had to be abandoned after 27 h, on 15 July 2012.

The collected passive-acoustic data were very noisy, which prohibited a semi-automated analysis (application of harbor porpoise-specific detectors and classifiers) of the data set. The manual analysis was difficult; it took approximately 8 work days to thoroughly analyze the 27 h of collected acoustic data (73 GB total). Several noise sources were identified: mechanical noise generated by the rudder and sail motors (mainly solid-borne sound), waves splashing against the boat hull, and general surface activity such as breaking waves and rain.

Nevertheless, the manual acoustic data analysis revealed that during the 27 h survey, 98 harbor porpoise click trains were registered. An example is shown in Fig. 2.

The registered click trains were comparatively short, each lasting between 1 and 3 s in duration. A map indicating the locations of the harbor porpoise acoustic encounters is shown in Fig. 3. About 10 % of the encounters were recorded during

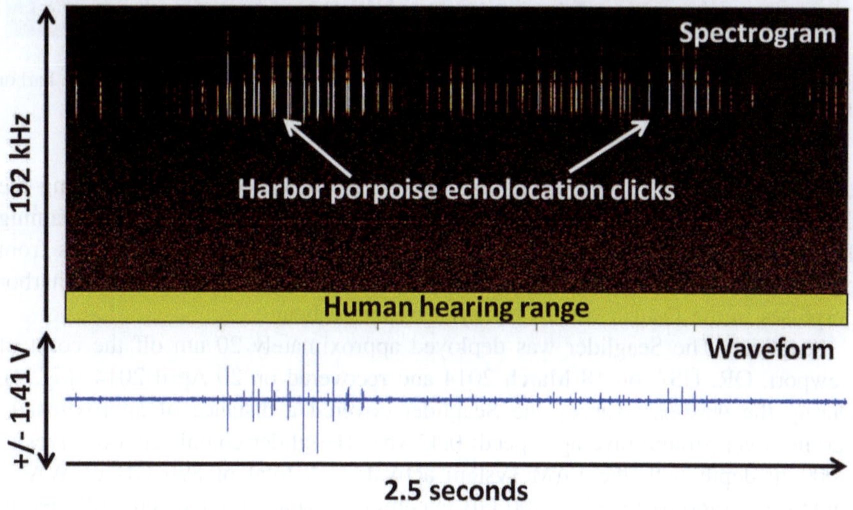

Fig. 2 Spectrogram and waveform of a harbor porpoise echolocation click train recorded with the Roboat. The *yellow box* in the spectrogram indicates the frequency range of human hearing (roughly 20 Hz–20 kHz). De-noising algorithms were applied to the data to eliminate electronic noise artifacts

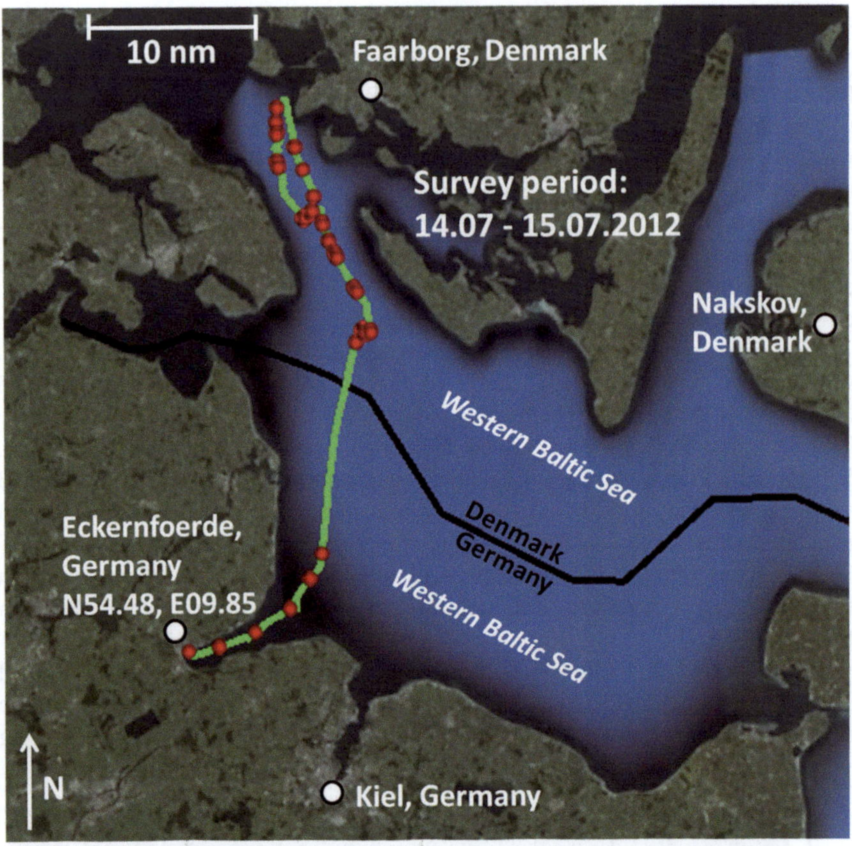

Fig. 3 Overview map of the Roboat survey. The *red dots* on the *green trackline* indicate harbor porpoise detections. Map: Google Earth

the first 4 h of the survey in German waters. No detections occurred during the following 4 h of the survey. Most encounters (90 %) occurred during the remaining 19 h in Danish waters north of 54.81°N. These observations match results from previous aerial surveys (e.g., [14]) which indicated a high abundance of harbor porpoises in the Danish waters surveyed by the Roboat.

Seaglider: The Seaglider was deployed approximately 20 nm off the coast of Newport, OR, USA on 18 March 2014 and recovered on 20 April 2014 (Fig. 4). During the west-east survey the Seaglider covered a distance of approximately 320 nm over ground (average speed: 0.42 kn). The glider completed 148 dives to 1,000 m depth with the PAM system activated. A total of 896 GB of WAC0 compressed (approximately 1,800 GB uncompressed) audio data were collected in the 25–1,000 m depth range. Vehicle-related self-noise was minimal and limited to times when glider-internal control and steering mechanisms (buoyancy pump, etc.)

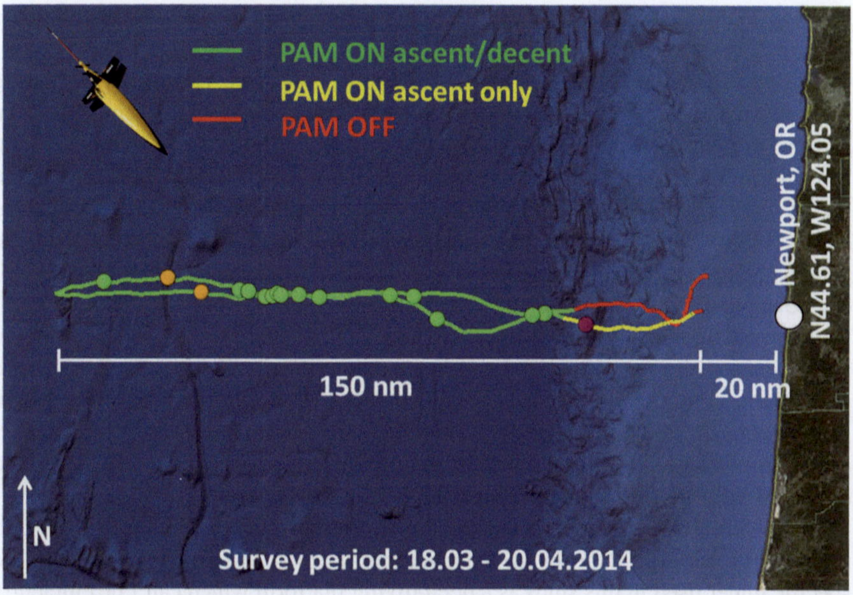

Fig. 4 Overview map of the Seaglider survey. The *green dots* on the trackline indicate Dall's porpoise detections, *orange dots* potential *Kogia spp.* detections, and the *purple dot* a potential detection of a mixed harbor and Dall's porpoise group. Map: Google Earth

were activated. The associated data loss is typically on the order of less than 10 % of the total dive time, but differs from dive to dive.

The acoustic data were manually analyzed for echolocation clicks of harbor porpoises, Dall's porpoises, and dwarf and pygmy sperm whales which are common in the study area. These four species are known to produce high-frequency echolocation clicks with a peak frequency of around 130 kHz. However, because of the similarity in the acoustic characteristics of their echolocation clicks, identifying to the species level remains challenging (e.g., [8]).

High-frequency cetaceans were recorded in 20 of 148 dives (14 %). As indicated by the green marks in Fig. 4, the gilder most frequently registered vocalizations produced by Dall's porpoise. The average acoustic encounter duration was 3 min.

4 Discussion

The different deployment scenarios did not allow a direct performance comparison between the two platforms. There were also no stationary passive acoustic recorders deployed concurrently which might have provided further insights into the effectiveness of the tested systems. However, the primary goal of this study was to evaluate the capabilities of mobile autonomous platforms (AUVs and USVs) to monitor high-frequency cetaceans producing echolocation clicks at frequencies beyond 100 kHz.

Roboat: The acoustic data recorded with the Roboat were very noisy. For future surveys one or more hydrophones should be towed at some distance behind the boat and at a greater depth. This will help to reduce both surface-induced and boat-induced noise. One of the advantages of USVs is that they can be operated in (very) shallow water, which is especially important in the context of tidal and wave energy efforts. Furthermore, some USVs are capable of moving faster than the animals being monitored and consequently standard distance sampling methods [3] can be applied to derive animal densities. Disadvantages include the 'liability issue' when operating the boat autonomously in coastal near-shore waters, where potential interference with recreational and commercial activities is likely. Also, USVs, and particularly autonomous sailboats, have not been used extensively for long-term passive-acoustic monitoring efforts. Thus more research and development is necessary to evaluate the full potential of these platforms, particularly with regard to effects of weather on monitoring ability.

Seaglider: The Seaglider persisted throughout the deployment and collected high-quality acoustic data for an extended period of time (almost 1 month). The specific glider used in this study is a deep-diving platform which can't be efficiently operated in shallow water. For this reason, this platform is most useful to monitor deeper, offshore areas, such as monitoring in conjunction with the installation and operation of offshore floating wind turbines. However, other types of commercially-available gliders are better suited to monitor coastal inshore areas. Once deployed, AUVs can be operated in most weather conditions and don't pose any navigational hazard. One of the disadvantages of AUVs, and particularly gliders, is that they move slowly through the water column (max. 0.5 kn) and therefore have difficulties dealing with strong currents.

Both the Roboat (in the Baltic Sea) and the Seaglider (in the North Pacific) successfully registered these transient signals and exemplified the potential of these platforms to be used for passive-acoustic monitoring efforts. This was the first time high-frequency echolocation clicks have been recorded using these platforms. In fact, this is the first report of a successful at-sea trial to acoustically monitor any marine mammal vocalizations using an autonomous sailboat.

Because of the limited detection range of the high-frequency echolocation clicks (a few 100 m), moving platforms are more effective in scanning and monitoring areas of interest than stationary recording devices. This is especially true for areas of low animal density (e.g., harbor porpoises in the eastern Baltic Sea).

Acknowledgments The authors would like to thank the sailing club in Eckernfoerde, Germany and especially Hans-Juergen Wurm for their assistance with the Baltic Sea field trials. We furthermore thank Zen Kurokawa and Sorqan Chang-Gilhooly for their help implementing the high-frequency PAM system into the Seaglider. Also thanks to Niki Diogou, Amanda Holdman, Jay Peterson, Ken Serven and the crews of the R/V Pacific Storm and R/V Elakha for assisting in the deployment and recovery of the Seaglider. This project was partially funded by the U.S.

Department of Energy, the U.S. Navy's Office of Naval Research, and NOAA. Co-author Selene Fregosi was supported by the U.S. Department of Defense through the National Defense Science and Engineering Graduate Fellowship (NDSEG) Program.

References

1. Bailey H, Brookes KL, Thompson PM (2014) Assessing environmental impacts of offshore wind farms: lessons learned and recommendations for the future. Aquat Biosyst 10:8
2. Baumgartner MF, Fratantoni DM, Hurst TP, Brown MW, Cole TVN, Van Parijs SM, Johnson M (2013) Real-time reporting of baleen whale passive acoustic detections from ocean gliders. J Acoust Soc Am 134:1814–1823
3. Buckland ST, Anderson DR, Burnham KP, Laake JL, Borchers DL, Thomas L (2001) Introduction to distance sampling. Oxford University Press, Oxford 448 pp
4. Cada G, Ahlgrimm J, Bahleda M, Bigford T, Stavrakas SD, Hall D, Moursund R, Sale M (2007) Potential impacts of hydrokinetic and wave energy conversion technologies on aquatic environments. Fisheries 32:174–181
5. European Wind Energy Association (2015). http://www.ewea.org/statistics/offshore-statistics/
6. Klinck H, Mellinger DK, Klinck K, Bogue NM, Luby JC, Jump WA, Shilling G, Litchendorf BT, Wood AS, Schorr GS, Baird RW (2012) Near-real-time acoustic monitoring of beaked whales and other cetaceans using a Seaglider™. PLoS ONE 7(5):e36128
7. Klinck H, Stelzer R, Jafarmadar K, Mellinger DK (2009) AAS endurance—an autonomous acoustic sailboat for marine mammal research. In: Proceedings of the international robotic sailing conference, Matosinhos, Portugal, July 2009, pp 43–48
8. Kyhn LA, Tougaard J, Breedholm C, Jensen FH, Ashe E, Williams R, Madsen PT (2013) Clicking in a killer whale habitat: narrow-band, high-frequency biosonar clicks of harbour porpoise (*Phocoena phocoena*) and Dall's porpoise (*Phocoenoides dalli*). PLoS ONE 8(5): e63763
9. Kyhn LA, Tougaard J, Thomas L, Duve LR, Stenback J, Amundin M, Desportes G, Teilmann J (2012) From echolocation clicks to animal density—acoustic sampling of harbor porpoises with static dataloggers. J Acoust Soc Am 131:550–560
10. Leung DYC, Yang Y (2012) Wind energy development and its environmental impact: a review. Renew Sustain Energy Rev 16:1031–1039
11. Mead JG, Brownell RL Jr (2005) Order cetacea. In: WilsonDE, Reeder DM (eds) Mammal species of the world (3rd ed.). Johns Hopkins University Press, Baltimore, 142 pp
12. Northwest National Marine Renewable Energy Center (2015). http://nnmrec.oregonstate.edu/
13. Rudnick DL, Davis RE, Eriksen CC, Fratantoni DM, Perry MJ (2003) Underwater gliders for ocean research. Mar Technol Soc J 38:73–84
14. Scheidat M, Gilles A, Kock K-H, Siebert U (2008) Harbour porpoise *Phocoena phocoena* abundance in the southwestern Baltic Sea. Endang Species Res 5:215–223
15. Stelzer R, Jafarmadar K (2012) The robotic sailing boat ASV roboat as a maritime research platform. In: Proceedings of 22nd international HISWA symposium on yacht design and yacht construction, Amsterdam, Netherlands. http://www.hiswasymposium.com/assets/files/pdf/2012/Stelzer.pdf
16. Tougaard J, Wright AJ, Madsen PT (2015) Cetacean noise criteria revisited in the light of proposed exposure limits for harbor porpoises. Marine Poll Bull 90:196–208
17. Tougaard J, Carstensen J, Teilmann J, Skov H, Rasmussen P (2009) Pile driving zone of responsiveness extends beyond 20 km for harbor porpoises (*Phocoena phocoena*). J Acoust Soc Am 126:11–14
18. Villadsgaard A, Wahlberg M, Tougaard J (2007) Echolocation signals of wild harbor porpoise, *Phoceona phoceona*. J Exp Biol 210:56–64
19. Willis PM, Baird RW (1998) Status of the dwarf sperm whale, *Kogia simus*, with special reference to Canada. Can Field-Nat 112:114–125

Design and Development of a Self-Stabilizing, Autonomous Sailboat with Zero-Net Stored-Energy Use

Robert Baker, Laura Kambourian, Sohrab Hajarian, Thomas Augenstein, Stephen Harnett, Gyu-Min Lee, Mukund Sudarshan, Cole Richter, Colette Trouillot, Philippe Williamson and Andy Ruina

Abstract We are developing a robotic boat capable of sailing semi-autonomously for two years on the open ocean. Our design is a monohull sailboat with a weighted keel, no rudder, a controlled-angle sail, and an air-rudder to passively control the boats orientation relative to the wind without active control. We are working to optimize sail, keel and air-rudder parameters to maximize directional stability and forward speed. Given the boats polar plot (speed vs. angle to the wind) we have begun the development of navigation schemes that utilize the ocean currents to travel more quickly and avoid obstacles. The boat controls will be powered by solar panels mounted on the deck. Taking latitude, season and cloud cover into account, the boat will need to function with a daily-average power budget of about 0.4 W. Contin-

---

R. Baker (✉) · L. Kambourian · S. Hajarian · T. Augenstein · S. Harnett · G.-M. Lee · M. Sudarshan · C. Richter · C. Trouillot · P. Williamson · A. Ruina
Autonomous Sailboat Team, Cornell University, Upson Hall 247, Ithaca, NY 14850, USA
e-mail: rwb266@cornell.edu

L. Kambourian
e-mail: lkk34@cornell.edu

S. Hajarian
e-mail: sh637@cornell.edu

T. Augenstein
e-mail: tea43@cornell.edu

G.-M. Lee
e-mail: gcl54@cornell.edu

M. Sudarshan
e-mail: ms2666@cornell.edu

C. Richter
e-mail: car245@cornell.edu

C. Trouillot
e-mail: cet54@cornell.edu

P. Williamson
e-mail: ppw22@cornell.edu

A. Ruina
e-mail: ruina@cornell.edu

A. Friebe and F. Haug (eds.), *Robotic Sailing 2015*,
DOI 10.1007/978-3-319-23335-2_4

uous use of the major electrical components in the sailboat requires much more than 0.4 W: the sail and air rudder servos consume 1.2 W in full capacity mode together, the Arduino Mega micro-processor consumes 0.84 W, and the GPS consumes 0.15 W. Thus, we can only use these components for brief time intervals, leaving them asleep or off for most of the time. One past iteration of the boat incorporated a freely rotating sail and tail to produce an angle of attack that generated lift for forwards propulsion, with an active water rudder for steering. We will test other self-stabilizing designs by interchanging parts to allow (1) Free or angle-controlled main sail; (2) tail angle controlled by servo or by a passive mechanism; (3) The presence or absense of an air rudder on the boat in addition to the tail; (4) The presense or total absense of a traditional water rudder.

1 Introduction

Environmental monitoring of the oceans is expensive. Our goal is to design a cheap small boat that, if produced in quantities of thousands, would greatly decrease the cost of various simple environmental data such as ocean temperature, salinity, acidity, cloud cover, etc.

This paper describes our considered designs as well as our ideal future design. We now have a monohull sailboat with a freely rotating sail, mechanically controlled air tail to control angle of attack, and a traditional water rudder. Later, we will switch to using a controlled-angle sail, mechanical tail and no water rudder.

2 State of the Art

Various models for autonomous sailboats have been built and tested by others. They are mostly motivted by a general need for environmental surveillance of the seas [1]. Here are a few.

Saildrone is an autonomous sailboat with a freely rotating sail with controlled tail and an actively controlled water rudder [2]. Its length is 5.8 m and has a 6.10 m^2 sail. It can go 1.54–2.57 m/s and carry a 90.7 kg payload. It uses 5–10 W of electrical power from solar cells [3]. Saildrone has navigated 8050 Km at sea [4].

Like Saildrone, uses a freely rotating sail and controlled tail [2]. But, to lower costs, our boat is smaller, at just 1 m long. Compared to Saildrone, our boat has about 1/30 the available solar panel area and 1/30 the available electical power. Thus, our design must function with 1/30 the power consumption.

FASt is an autonomous sailboat 2.5 m long with a rig 2.6 m high [5]. FASt uses two servos to control twin rudders, and one motor to control both sails. As for essentially all long-distance robotic sailboats, navigation is based on data from GPS, a tilt-compensated compass, and wind-speed and direction sensors, all using electricity fom solar-panel recharged batteries.

While FASt uses a traditional sailing rig to navigate, one of our considered designs uses, like Saildrone, a freely rotating sail shaped like a wing. And our planned design uses the controlled wing and tail for steering.

The U.S Naval Academy's autonomous sailboat team wants to complete a trans-Atlantic trip [6]. Their keel has a 30-degree sweep angle to avoid tangling with debris (such as seaweed or floating lines). Their design uses an Airmar WeatherStation PB200 sensor, which provides GPS, heading and true and apparent wind direction data, and linear actuators for sail and rudder position control [7].

The electronics of our boat are similar to those of the Naval Academy's boat.

Our boat Our long-term planned design addresses an issue not addressed in the designs above, achieving directional stability with low energy use. Long distance sailing with just 0.4 watts of available electrical power demands this. Some human carrying boats made for small-crew long-distance sailing have auto-steering systems that automatically steer relative to the wind. These are complex contraptions. Our planned design is a simple solution to the same problem. Essentially our design is to skip the water rudder and to make the whole sailing rig a weather vane that the boat follows. Boat heading control is by turning the boat under this vane (the controlled sail angle).

3 Boat Design

3.1 Mechanisms

One of our designs uses a freely-rotating sail, mechanical tail, hull, keel, and water rudder are shown schematically in Fig. 1.

A recent prototype uses a freely-rotating symmetrical foam-core airfoil. This sails angle with respect to the wind is controlled by a mechanism fixed to the base of the sail that controls an auxiliary airfoil, or tail, positioned behind the main sail. The tail generates a moment on the sail to create a desired angle of attack. This mechanism eliminates the need for a servo to control the tail, saving electrical power. The hull possesses a nearly semicircular cross-sectional shape with a thin, symmetrical airfoil-shaped keel.

The low center of mass was used to create roll stability. The keel is made of steel and weighs 4 kg. The disparity in mass distribution results in a self-righting body. The keel has a steel ballast to increase this righting torque. The water rudder is positioned at the stern of the boat, controlled by a servo in a housing that hangs off the back of the deck.

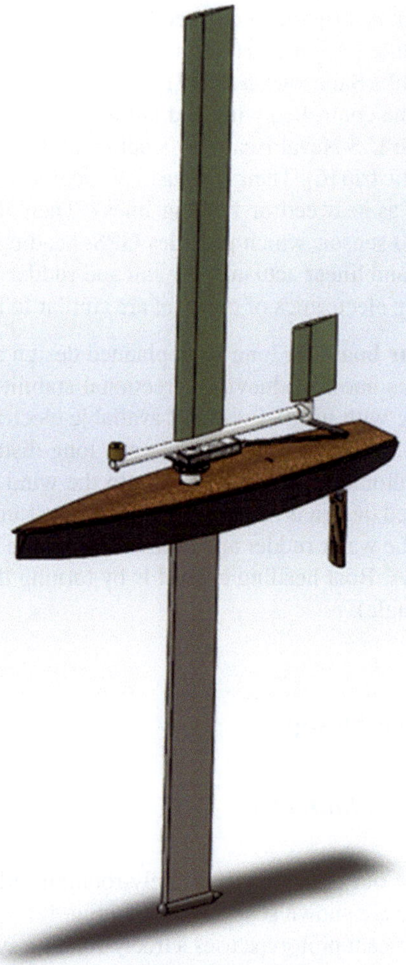

Fig. 1 Schematic CAD Model for the previous boat. This version has a free sail with angle of attack governed by the controlled tail. Steering is with a water rudder

3.2 Hull Constructions

The hull needs to be durable, easy to build, easy to control, easy to model, and capable of decent speed and stability on fresh or seawater. The initial design is based on the CAD surface of an International One Meter model racing boat shown in Fig. 2 [8]. After adjusting the surface and making dependent sketches, a model was lofted on which to base construction. Several actual-size top-down cross section sketches were printed and used to cut out one inch thick foam sections. These sections were glued together and the corners smoothed out using a Surform tool. With the foam core carved and sanded, two layers of fiberglass and an outer layer of primer were applied and sanded.

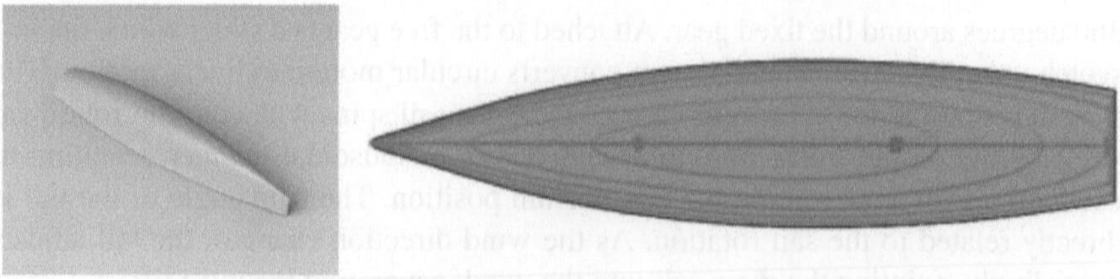

Fig. 2 CAD model of boat hull to be foam cut

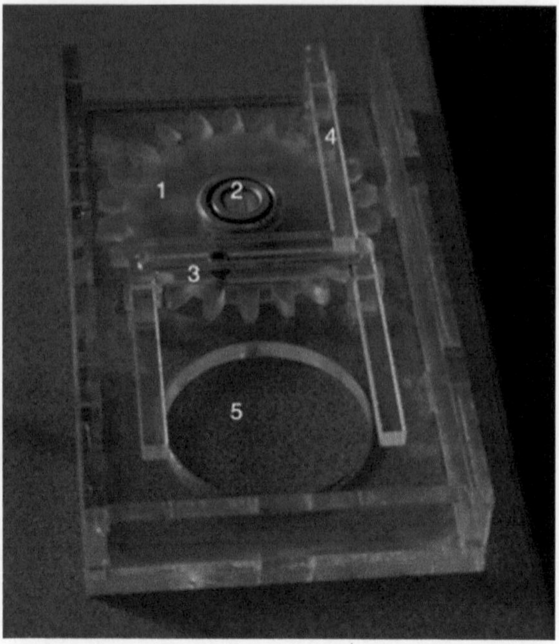

Fig. 3 Image of Auto-Trim System when not connected to the sailboat. *1*-FreeGear. *2*-Axle fixed in box that free gear rotates on. *3*-Slider part of the scotch yoke attached to free gear. *4*-Slider that pulls on an arm (not shown) attached to the trimmed tail. *5*-Hole for axle mount. The axle slides through this hole and fixed gear attaches rigidly to axle mount

3.3 Sail and Auto-Trim System Construction

The sail was constructed out of foam, using a hand-built foam cutter and an airfoil-shaped guide. The sail weighs 0.54 kg and is 1.30 m tall. It rotates about a carbon fiber tube axle which is mounted on bearings supported by the deck. A layer of fiberglass was applied to the sail for rigidity and durability.

The sail has to be directed at an angle of attack from the apparent wind direction to generate lift forward, which is accomplished using a tail [22]. The auto-trim system (Fig. 3) rotates the tail in the appropriate direction with only mechanical linkages, preserving electrical energy compared to using a motor to turn the tail.

The auto-trim system controls the tail which is attached to our main sail. The tail is attached at a set distance away creating a moment about the sail. One gear is rigidly attached to the sail mount while another gear, housed in a box, is free to rotate the full

360 degrees around the fixed gear. Attached to the free gear is a slider which fits in a scotch yoke; this mechanical system converts circular motion to linear motion. The system is optimal for tail trimming because if the sail spins with constant rotational velocity, the output linear motion accelerates in a sinusoidal manner, resulting in oscillating trim angles about the equilibrium position. The trim angle of the tail is directly related to the sail rotation. As the wind direction changes, the tail adjusts accordingly, pointing the main sail into the wind, generating the most lift.

3.4 Electrical Tail Option

Alternatively, an electrical tail can trim the sail by connecting a servo motor to a hinged tail. The advantage of this method is direct control over the tail angle, allowing two benefits over the auto-trim system: (1) A constant tail angle, as opposed to one that varies with sail angle, reaching a maximum when the sail is perpendicular to the boat and a minimum when parallel and (2) Ability to alter angle of attack, depending on wind speed.

3.5 Keel Construction

To provide roll stability, the boat has a long, narrow keel made of steel with a steel ballast. The keel has the shape of a NACA 0005 airfoil made from a flat bar of steel that was cut and grinded [9]. The ballast was created with the shape of a torpedo as a model, using a solid steel rod with a 3D-printed nose and tail cap.

3.6 Control Surface Concepts

The boat is a testing platform for four control surface concepts: Controlled sail and tail, controlled sail and air rudder, free sail with controlled tail and air rudder, and an auto-steering system.

Figure 4 shows the first three concepts. Each design uses a tail or air rudder to set the heading of the boat based on the direction of the wind. In each design there are control surfaces that will orient themselves at a certain angle to the direction of the wind. If the control surface is fixed to the boat it will alter the heading of the boat when it orients itself. The largest black outline represents the hull of the boat and the smaller blue outlines are airfoils. The solid black form in the middle of the boat represents the position of the keel.

First, Fig. 4a shows the forced sail steering system. The lift generation sail at point 1 is fixed. The sail at point 1 and the tail at point 2 are actuated with servos. The angle of attack for the sail is produced by the actuated tail. The sail is no longer freely

Fig. 4 Clockwise from *top left* Figs. a, b and c

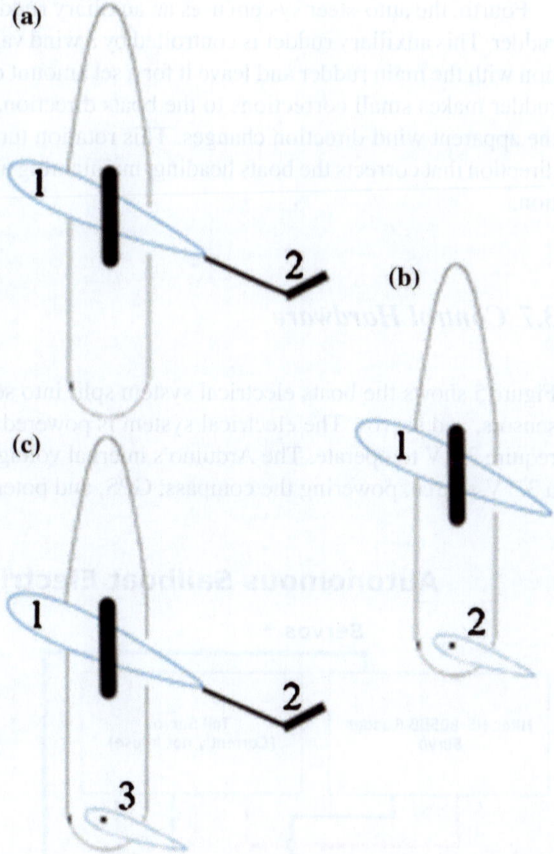

rotating, the sail will reorient the boat when it takes this angle of attack. By having a servo on both control surfaces at point 1 and point 2, the heading of the boat can be chosen relative to the direction of the wind. This design is advantageous because it reduces the number of control surfaces and the number of moving parts.

Second, Fig. 4b shows a freely rotating sail at point 1 and tail at point 2, actuated with servos. The servo at point 1 is used to create an angle of attack of the sail. The servo at point 2 is used to set the heading of the boat because this airfoil will align itself with the wind, taking a zero degree angle of attack, and due to its long moment arm about the keel, it will exert a moment on the boat that alters the heading.

Third, Fig. 4c shows a combination of the first two options: the freely rotating, lift-generating sail at point 1 with actuated tail at point 2 and actuated air rudder at point 3. The servo at point 2 is used to angle the tail which will force the sail to orient itself with some angle of attack since the sail can freely rotate. The servo at point 2 is used to set the heading of the boat because this airfoil will align itself with the wind, taking an angle of attack of zero.

Fourth, the auto-steer system uses an auxiliary rudder positioned behind the main rudder. This auxiliary rudder is controlled by a wind vane. The idea is to set the direction with the main rudder and leave it for a set amount of time, knowing the auxiliary rudder makes small corrections to the boats direction. The wind vane rotates when the apparent wind direction changes. This rotation turns the auxiliary rudder in the direction that corrects the boats heading, maintaining a constant apparent wind direction.

3.7 Control Hardware

Figure 5 shows the boats electrical system split into sections for processing, power, sensors, and servos. The electrical system is powered by 6 V batteries. The sensors require 3.3 V to operate. The Arduino's internal voltage regulator is used to produce a 3.3 V source, powering the compass, GPS, and potentiometer.

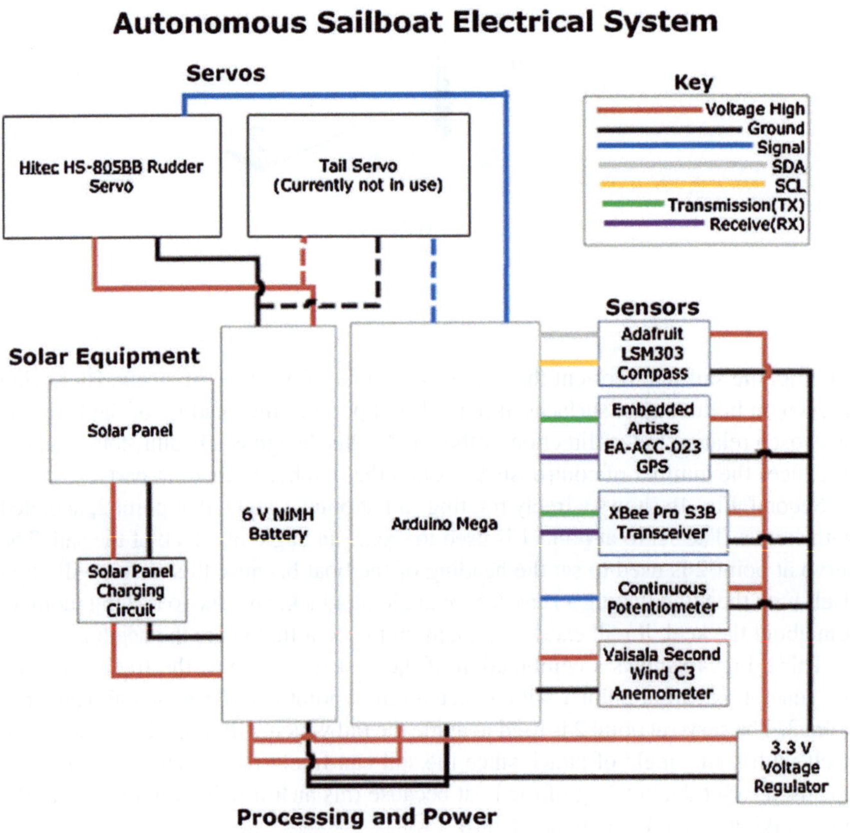

Fig. 5 Electrical system schematic

The compass is a magnetometer-accelerometer combination. The accelerometer is used to calculate the compasss orientation (pitch and roll). This allows the addition of tilt compensation to the heading. The GPS unit is used to determine the boats location and velocity. In the freely rotating sail model, a continuous rotation potentiometer is attached to the bottom of the mast to determine its angle and therefore calculate wind direction. The anemometer provides wind speed and is mounted on the bow of the boat. An XBee Pro S3B transceiver sends data from the Arduino to a computer on land for debugging and manual rudder control. The dashed lines in the diagram represent the connection to a tail servo. The controlled tail is an alternate design choice for the mechanical tail.

3.8 Energy Harvesting

Energy harvesting analysis and budgeting applies to long-duration sea voyages. This analysis computes the amount of time the electronics can be in full-capacity mode versus sleep mode based on amount of energy the solar panels can harvest.

Various electrical components, such as the microcontroller and motors, will use power from a battery pack. This battery pack will be recharged via solar panels mounted on the top surface of the deck. The following formula calculates the amount of energy that can be harvested via the solar panels [11]:

$$E = CF * TF * PGT * R_{panel} * t * A \tag{1}$$

where E = harvested energy (W*h); CF = Cloud Factor (0.1); TF = Tilt Factor (0.6); PGT = Plexiglass transmissivity (0.9) [12]; Rpanel = power rating of the panel (147.84 W/m^2); t = length of time (h); and A = surface area (0.2 m^2)

The cloud factor takes into consideration the decrease in solar insolation on the PV cells as a result of clouds being present, which is at maximum 90 %. The tilt factor incorporates the efficiency losses for mounting the solar panels flat on the deck rather than at an optimal angle. The equation which accounts for transmissitivity losses as light passes through the plexiglass incorporates values for plexiglass transmissivity. The usable surface area on the deck of the boat is 0.2 m^2.

This calculation is sensitive to two factors: season and latitude. Table 1 below presents information on operating conditions to calculate harvestable power. Bold italic and bold numbers represents the maximum and minimum respectively.

3.9 Energy Budget

The power consumption of the electrical components for full capacity and sleep mode are provided in Table 2 below.

Table 1 Power available as dependent on latitude and season

Latitude (N°) [13, 14]	Date	Length of day (h)	Average solar insolation (W/m^2) [15]	Energy harvested (W*h)	Average power (Energy/24 h)
0	January 1	12	500	17.7408	0.7932
	April 1	12	600	17.7408	0.7932
	July 1	12	650	17.7408	0.7932
	October 1	12	620	17.7408	0.7932
10	January 1	11.43	490	16.8991	0.7041
	April 1	12.09	580	17.8739	0.7447
	July 1	12.58	620	18.5983	0.7749
	October 1	11.91	590	17.6077	0.6659
20	January 1	10.81	450	15.9815	0.6659
	April 1	12.18	550	18.0069	0.7503
	July 1	13.20	590	19.5149	0.8131
	October 1	11.82	560	17.4747	0.7281
30	January 1	10.10	420	14.9318	0.6221
	April 1	12.28	500	18,1548	0.7564
	July 1	13.91	530	20.5645	0.8568
	October 1	11.58	510	17.1199	0.7022
40	January 1	9.21	400	13.6161	0.5673
	April 1	12.41	500	18.3469	0.7644
	July 1	15.81	520	21.8951	0.9123
	October 1	11.58	510	17.1199	0.7133
50	January 1	7.93	300	11.7237	0.4884
	April 1	12.58	420	18.5983	0.7749
	July 1	16.09	500	23.7875	0.9911
	October 1	11.40	430	16.8538	0.7022
60	January 1	5.66	200	**8.3677**	**0.3486**
	April 1	12.84	410	18.9827	0.7909
	July 1	18.38	500	***27.1730***	***1.1322***
	October 1	11.13	430	16.4546	0.6856

For long-term navigation, the integrated system of the microcontroller, sensors, and actuators needs to be designed to function with the harvestible amount of electrical energy. To accomplish this, we needed to spread our available electrical energy throughout the day by periodically turning off various electrical components. Consequently route corrections would be made periodically rather than continuously. The only components left on during the boats sleep mode will be the microcontroller and the GPS, which will consume 0.1492 W together. The transceiver will be turned off during the boats sleep mode.

Table 2 Power budget specifications in by component

Electrical component	Power consumption—full capacity (W)	Power consumption—sleep mode (W)
Rudder servo [16]	1.2	0
Tail servo [16]	1.2	0
Arduino mega 2530 [10]	0.84	0.08
Transceiver [17]	1.1	0.09
GPS [18]	0.148	0.01
Compass [19]	0.00033	0
Potentiometer [20]	0.00002	0
Total	4.48	0.18

A time factor was calculated below using a ratio between the times in full operation mode to time in sleep mode.

$$FC * t + SM * (24 - t)P * T_d \qquad (2)$$

Where FC = power consumption at full capacity (4.48 W); t = length of time electrical components operate (h); SM = power consumption during sleep mode (0.18 W); P = power output of solar panels, (minimum: 0.35 W); Td = length of sunshine in a day, (minimum: 5.66 h or 2.04E4 s)

Solving for *t*, the minimum total time the entire system can operate is 0.46 h, producing a time factor of 52.

3.10 Qualitative Wet Trials

Pool testing consisted of qualitative analysis of the boats stability, waterproofing, lift generation, and RC system. Since the design creates stability from a ballast, weight was added incrementally to the bottom of the keel to determine an appropriately weighted ballast (as light as possible while producing roll stability). To test lift generation, a setup of two vertically stacked fans on a roller was dragged alongside the pool to create a wind field around the boat. The RC system was tested by giving commands to the boat tail and rudder servos. The main advantages of testing this in the pool instead of dry testing was to observe the mechanical effectiveness of the rudder, estimate appropriate rudder angles, and estimate appropriate tail angles to generate lift effectively [21].

4 Boat Dynamics

4.1 Simulations

A dynamic simulation of a sailboat was created based on a 2D model of a sailboat which treated the sailboat as a sail, keel, rudder, and hull. A MATLAB simulation was created to predict the trajectory of a sailboat under various conditions. The simulation uses MATLABs ode45 function to numerically solve the equations of motion of the boat over time. A MATLAB GUI was created to allow users to easily modify parameters, initial conditions of the boat, and other options.

The GUI allows the user to input the initial pose and velocity of the boat in the Initial Conditions section. The Misc. Parameters section allows control over various parameters such as wind velocity and direction, boat mass, boat moment of inertia, water density, and air density. The current simulation allows only for a uniform wind velocity field and assumes water velocity to be zero. The sail, keel, and rudder parameters can also be selected. In the sail and rudder sections, there is also a drop down menu to select different types of sails and rudders. In the Animation section, the duration and frames per second of the animation can be chosen. A drop down menu also allows the user to choose between a global fixed view of the boats trajectory and a boat centric view. Under the Lift/Drag Accuracy section, a PCHIP interpolation or a sinusoidal approximation can be selected as the method used to estimate the lift and drag coefficients for a given angle of attack. In the Active Control section, the user has the option of controlling the sail angle, rudder angle, or both angles actively during the simulation using the arrow keys on the keyboard. Once the desired options have been set, the user can click the Start button to run the simulation (Fig. 6).

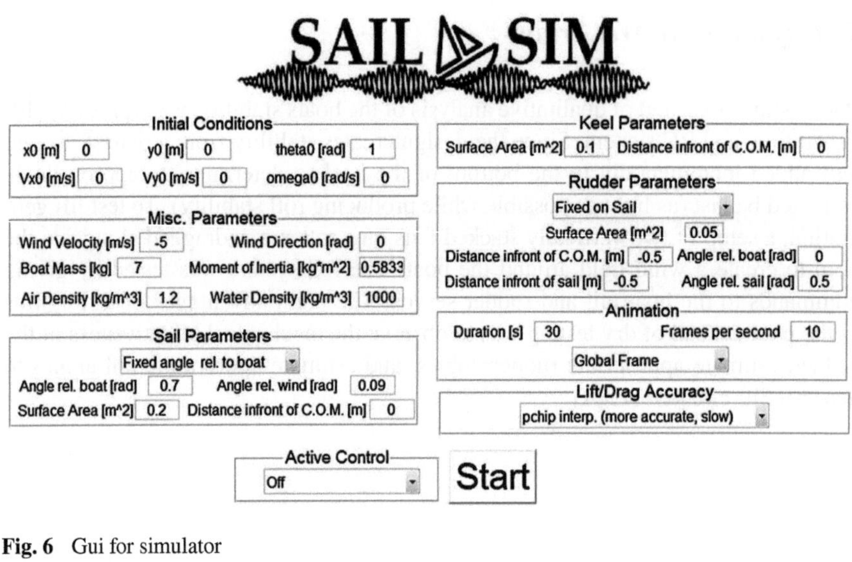

Fig. 6 Gui for simulator

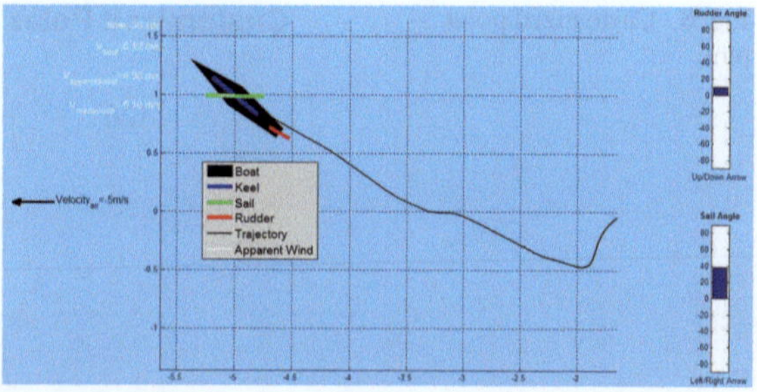

Fig. 7 Dynamics simulation

Figure 7 shows a screenshot of the animation from the simulation. The boats hull, sail, keel, and rudder are shown and labeled in the legend. The trajectory of the boat is also plotted. The runtime, velocity magnitude of the boat, velocity magnitude of the apparent wind, and velocity made good are displayed in the upper left corner. A small white line on the boat indicates the direction of the apparent wind. An arrow on the left side of the figure indicates the direction of the true wind. If active control is selected, the rudder and sail angles will be displayed on the right side and can be controlled with the arrow keys. After the animation is finished or the figure is closed, various plots showing the pose and velocity of the boat over time are plotted.

The speed characteristics of a given sailboat can be displayed via a polar plot [22]. The curve of a polar plot displays the expected speed of a boat at a given heading in a specified wind field, used to determine the preferred heading angles relative to the wind. As such, navigation can utilize the polar plot to optimize the path for the boat.

4.2 Polar Plot

To create a custom polar plot for the boat, the lift and drag of the sail and keel were used to determine the forward and lateral forces on the boat at every angle [23]. The polar plot (Fig. 8) displays boat speed when it has reached steady state (constant velocity), so these forward and lateral forces were matched to forward and lateral drag forces on the boat. From forward and lateral tow tests in the pool, a relationship between drag force and speed was determined, allowing computation of the speed at each angle, similar to tests done with other autonomous sailboat prototypes [24].

5 Navigation

5.1 Short Term Path Planning

The short-term navigation algorithm determines the optimal boat heading for reaching a target destination. As inputs, the algorithm takes in water current position, water current heading, wind direction, target destination, and the boat polar plot. The output from the algorithm is the new desired boat heading. The algorithm finds the

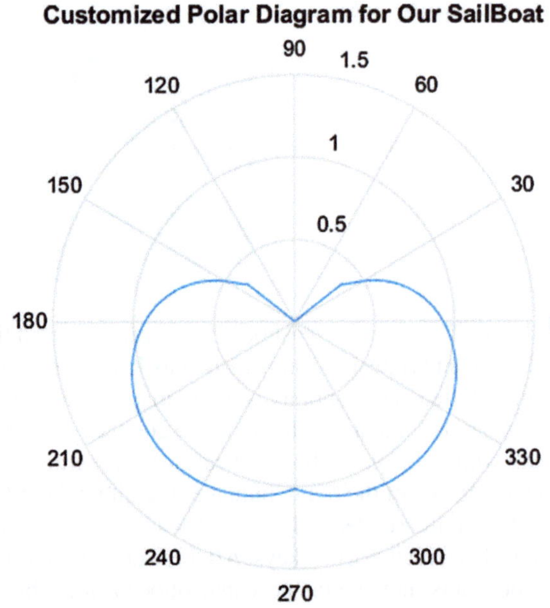

Fig. 8 Customized polar plot

desired heading by using the boat polar plot to find the heading that will maximize the velocity towards the target destination. Water current direction and magnitude are not taken into account. The algorithm can be seen below, with variables depicted in Fig. 9.

For the MATLAB code, the inputs used were Pose_{Boat}, Target, θ_{wind}, polar plot. The outputs given by the MATLAB code is $\theta_{Desired}$.

$$r = Target - Pose_{Boat}[1,2] \tag{3}$$

$$n = 1 + P_c/(norm(r)) \tag{4}$$

$$v_B = [v_{B,mag}(\alpha) * cos(\theta_{wind} + \alpha), v_{B,mag}(\alpha) * sin(\theta_{wind} + \alpha)] \tag{5}$$

$$v_{T,R} = v_B * r/(norm(r)) \tag{6}$$

$$\alpha = \alpha + \Delta\alpha \tag{7}$$

The first step for the MATLAB code is to calculate distance vector from the boat to the target as shown in Eq. 3. The first two elements of Pose_{Boat} are the longitude and latitude of the boat. Afterwards, using Eq. 4 MATLAB calculates the scaling factor used to determine beating width of jibing and tacking maneuvers: where P_c is the parameter tuned to manipulate the beating width. A larger P_c results in a wider beat width.To find the boat angle that maximizes the velocity towards the target,

MATLAB iterates through the right half of the polar plot. MATLAB calculates the velocity of the boat at given angle α using Eq. 5. MATLAB iterates by running for all α 180 to 360, where α is an angle of the polar plot. Afterwards, MATLAB calculates the velocity towards the target using the Eq. 6 and checks if $v_{T,R}$ is greater than $v_{T,R,Max}$. If it is MATLAB saves it as new $v_{T,R,Max}$ and saves $\theta_{B,R,Max}=\theta_{Wind}+\alpha$. For the next iteration, MATLAB increases α using Eq. 7. MATLAB repeats the iteration for all α 0 to 180 to find $\theta_{B,L,Max}$ and $v_{T,L,Max}$. It then compares $\theta_{B,L,Max}$ and $\theta_{B,R,Max}$ with θ_B to choose which heading is best using the if loop shown below.

Check if heading from right half of polar plot is closer to current heading
If $(\theta_{B,R,Max} - \theta_B) < (\theta_{B,L,Max} - \theta_B)$
Check if boat should jibe/tack
If $(v_{T,R,Max} * n < v_{T,L,Max})$
$\theta_{Best} = \theta_{B,L,Max}$
else
$\theta_{Best} = \theta_{B,R,Max}$
else
Check if boat should jibe/tack
If $(v_{T,L,Max} * n < v_{T,R,Max})$
$\theta_{Best} = \theta_{B,R,Max}$
else
$\theta_{Best} = \theta_{B,L,Max}$
Output θ_{Best}

Once the desired heading has been found a proportional derivative (PD) controller controls the rudder to steer the boat towards the desired direction. Figure 10 represents the block diagram of the short term navigation algorithm.

The wind direction can be seen in the top left corner of Fig. 11. The green line attached to the boat is the current desired direction as calculated by the short-term navigation algorithm. The different trajectories between waypoints are a result of the polar plot. The final trajectory has the boat going into the wind. Since the boat cannot travel directly up wind, the short-term navigation has it perform a series of loops called jibes that allows the bow of the boat to never cross the wind direction vector.

Figure 10 shows the block diagram of the Short Course Routing System. Target is the desired position of the boat, θW is the angle of the wind, θD is the desired heading of the boat, θe is the error between the desired heading and the actual heading, θR is the angle of the rudder, and θB is the heading of the boat.

Figure 11 is the result of a MATLAB simulation using the short term navigation to visit multiple waypoints, represented by red stars. The boat starts at the bottom left way point and visits the waypoints in counter-clockwise fashion, returning to the start waypoint. The final segment of the navigation course shows the boat jibing upwind to its final orientation shown by the green line.

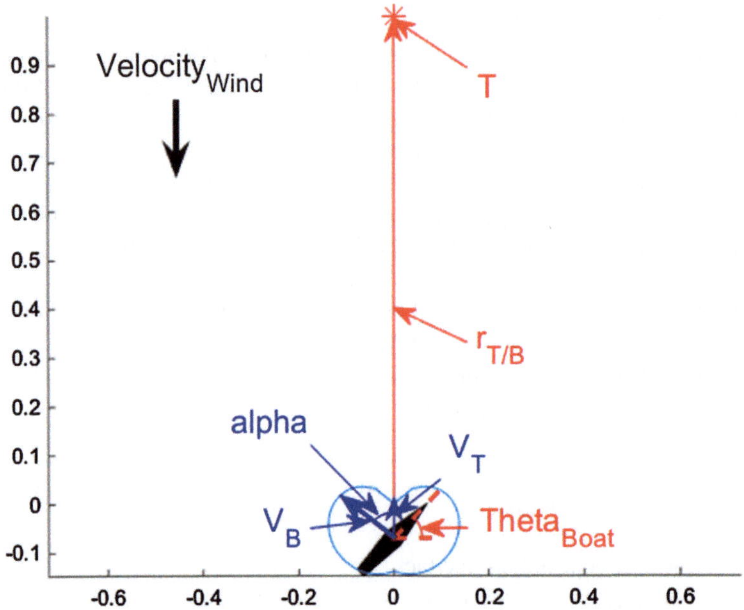

Fig. 9 Illustration of various variables used in short term navigation algorithm

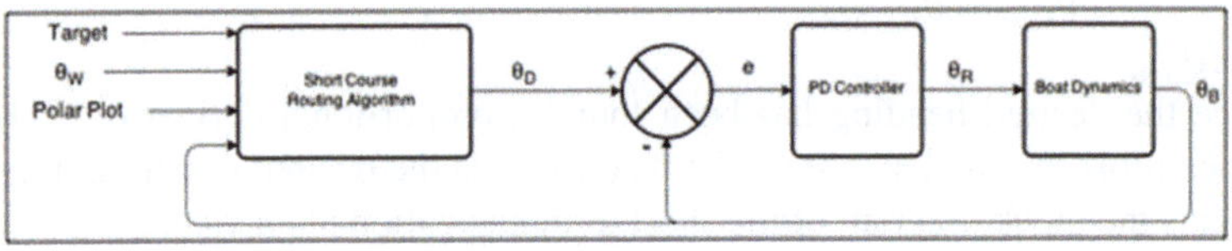

Fig. 10 Short course routing system Block Diagram

5.2 Long Term Path Planning

The following procedure describes how the program chooses an optimal trajectory for navigating long distances. The goal is to take into account the curvature of the earth, landmasses and weather. The basic idea is to use Dijkstras algorithm to find the shortest path, where shortest refers to the expected time taken to travel the distance between them [25].

The first step is to overlay a grid of nodes on the globe. These nodes are at intersections of latitude and longitude lines. The horizontal distance between the nodes is a multiple of a degree of latitude and the vertical distance between the nodes is a multiple of a degree of longitude. Nodes on land are then removed. Next, edge weights are calculated by passing in two nodes to our weight calculator function. This computes the straight-line distance between the nodes and checks if a landmass is between the nodes. If there is, it sets the weight to infinity.

Fig. 11 Matlab simulation

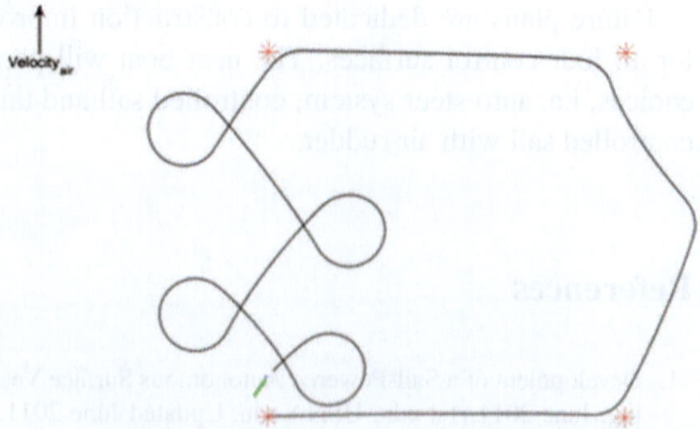

Fig. 12 Long term path planning from Houston to Tel Aviv

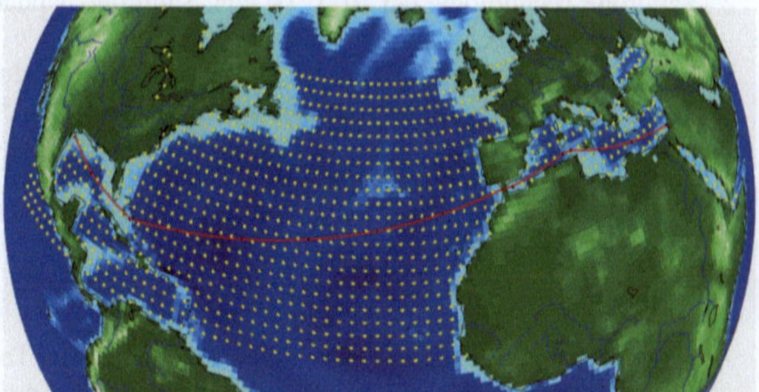

This information is stored in an adjacency matrix [26]. The final step is finding a node that is closest to the source and another that is closest to the destination. These nodes and the adjacency matrix are input into a separate function that uses Dijkstras algorithm to determine the shortest path. Figure 12 shows a test run where the source was Houston, Texas and the destination was Tel Aviv, Israel.

6 Conclusions

Our sailboat is in the iterative testing stage for both mechanical and electrical functionality. It is mechanically stable, waterproof, capable of turning, and capable of generating lift in the desired direction of motion. The sensors receive accurate data that is fed into the Arduino to send commands to the water rudder servo, although this process is not always reliable and requires troubleshooting. The short-term navigation algorithm is successful in simulations but is currently under testing in the field.

Future plans are dedicated to construction improvement and a testing platform for all four control surfaces. The next boat will permit testing of alternate design choices, i.e. auto steer system, controlled sail and tail, free sail with air rudder, and controlled sail with air rudder.

References

1. Development of a Sail-Powered Autonomous Surface Vessel (ASV) for Trans-Atlantic Voyaging. June 2011, 1st edn. USNA.edu. Updated June 2011. http://www.airmartechnology.com/uploads/brochures/pb200.pdf. Accessed June 2015
2. Saildrone Technology. Jan 2011, 1st edn. Saildrone. Updated Jan 2015. http://www.saildrone.com/index.php/technology. Accessed June 2015
3. Saildrone Capabilities. June 2011, 1st edn. USNA.edu. Updated Jan 2015. http://www.saildrone.com/index.php/capabilities. Accessed June 2015
4. 100 Days at Sea. June 2011, 1st edn. USNA.edu. Updated May 2013. http://www.saildrone.com/index.php/test/78-saildrone/88-100-days-at-sea. Accessed June 2015
5. FASt WRSC Team. 2014, 1st edn. WRSC 2014 Teams. Updated 2014. http://wrsc2014.com/teams/1. Accessed June 2015
6. Development of the USNA Sailbots (ASV). http://www.usna.edu/Users/naome/phmiller/IRSC-USNA.pdf. Accessed June 2015
7. Ultrasonic Weather Station Instrument. 1st edn. Airmar Technology. Updated Jan 2011. http://www.airmartechnology.com/uploads/brochures/pb200.pdf. Accessed June 2015
8. International One Meter Design. http://rcyachts.net/TS2/. Accessed June 2015
9. Airfoil Tools: NACA 0015 (naca0015-il). http://airfoiltools.com/airfoil/details?airfoil=naca0015-il. Accessed June 2015
10. Arduino Mega ADK (2012) Arduino. http://www.arduino.cc/en/Main/ArduinoBoardMegaADK?from=Main.ArduinoBoardADK. Accessed June 2015
11. Solar Radiation Data Manual for Flat-Plate (1994) NREL.gov. US Department of Energy. http://www.nrel.gov/docs/legosti/old/5607.pdf. Accessed June 2015
12. Plexiglass: General Information and Physical Properties (2006) Plexglass.com Altuglas international Arkema Group. http://www.plexiglas.com/export/sites/plexiglas/.content/medias/downloads/sheet-docs/plexiglas-general-information-and-physical-properties.pdf. Accessed June 2015
13. Duration of Daylight/Darkness Table for One Year (2011) The United States Naval Observatory. http://aa.usno.navy.mil/data/docs/Dur_OneYear.php. Accessed June 2015
14. Daylight Hours Explorer (2010) Astronomy Education at the University of Nebraska-Lincoln. http://astro.unl.edu/classaction/animations/coordsmotion/daylighthoursexplorer.html. Accessed June 2015
15. National Solar Radiation Data (2013) National Renewable Energy Laboratory. http://www.nrel.gov/disclaimer.html. Accessed June 2015
16. Parallax Standard Servo v2.2 (2010) Parallax Inc. https://www.parallax.com/sites/default/files/downloads/90000005StandardServoProductDocumentationv2.2.pdf. Accessed June 2015
17. XBee & XBee-Pro ZB Overview (1996) digi. http://www.digi.com/pdf/ds_xbeezbmodules.pdf. Accessed June 2015
18. GPS Standalone Module Data Sheet (2011) GolbalTop Technology Inc. http://www.embeddedartists.com/sites/default/files/support/acc/gps/GlobalTop-FGPMMOPA6H-Datasheet-V0A.pdf. Accessed June 2015
19. 3-Axis Digital Compass IC (2013) Honeywell. http://www.adafruit.com/datasheets/HMC5883L_3-Axis_Digital_Compass_IC.pdf. Accessed June 2015
20. Series 282 Data Sheet: Precision Potentiometer (2013) CTS Electrocomponents. http://www.ctscorp.com/components/Datasheets/282.pdf. Accessed June 2015

21. Aber Sailbot Autonomous Sailing Robotics Team. 29 Jan 2015, 1st edn. Aber Sailboat. Updated Jan 2015. http://abersailbot.co.uk/. Accessed April 2015
22. Polar Plot for Sailboats 11 Oct 2012, 1st edn. NauticEd. Updated Oct 2012. http://www.nauticed.org/sailing-blog/how-to-read-a-polar-plot-for-sailboats/. Accessed June 2015
23. Sailing 101: Understanding Polars Through Animation 22 Sept 2013, 1st edn. Updated Sept 2013. http://features.boats.com/boat-content/2013/09/sailing-101-understanding-polars-through-animation/. Accessed June 2015
24. Design of an Autonomous Sampling Boat for the Study of Algae Bloom in Lake Zurich. Oct 2009, 1st edn. Swiss Federal Institute of Technology Zurich. Updated Oct 2009. Retrieve at: Page 26. http://students.asl.ethz.ch/upl_pdf/167report.pdf. Accessed May 2015
25. A Note on Two Problems in Connexion with Graphs (2015) Dijkstra, E.W. http://www-m3.ma.tum.de/foswik/pub/MN0506/WebHome/dijkstra.pdf
26. Adjacency Matrices. 2015, 1st edn. UT Dallas. Updated 2015. http://www.utdallas.edu/jwz120030/Teaching/PastCoursesUMBC/M221HS06/ProjectFiles/Adjacency.pdf. Accessed May 2015

Kite Sailing Platform Mathematical Model and Stabilization

Konstantin Aprosin, Aleksander Tavlintcev, Sergey Semenenko and Maria Shorikova

Abstract This article is devoted to the mathematical modeling of a nonconventional sailing platform, which is called a mast-free (or kite) sailing platform. The platform is based on the modern sail type called "kite". It is a semi-rigid concave wing, which is used for towing water surface objects by wind power. Systems based on the kite successfully evolved over the past 10 years as an independent high-tech water sport. Mast-free sailing platforms managed by human allow them to travel long distances in a wide range of weather conditions. It is necessary to make the platform completely autonomous. To create an automatic control system for this new sailing platform it is necessary to have a mechanical model of the platform. This model should be linear in proximity of its equilibrium states for steady state calculations. All automatic stabilization theory is applicable to use within the linearized model.

1 Introduction

In recent years, active search of mobile robotic platforms is actual. One of perspective problems is oceanic autonomous sailing platform development. Nowadays, designers from Europe and USA attempt to create autonomous sail boat mostly according to classical single-sticker scheme. The alternative approach of sailing platform design is presented in the paper. It is called a mast-free (or kite) sailing

---

K. Aprosin (✉) · A. Tavlintcev · S. Semenenko · M. Shorikova
Ural Federal University, E-309, Mira street 19, Yekaterinburg, Russia
e-mail: aprosin.ki@gmail.com

A. Tavlintcev
e-mail: WINDDAES@gmail.com

S. Semenenko
e-mail: sinissem@gmail.com

M. Shorikova
e-mail: maria.morgenrote@gmail.com

A. Friebe and F. Haug (eds.), *Robotic Sailing 2015*,
DOI 10.1007/978-3-319-23335-2_5

platform. The platform is based on the modern sail type called "kite". It is a semi-rigid concave wing, which is used for towing water surface objects by wind power. Systems based on the kite successfully evolved over the past 10 years as an independent high-tech water sport [1]. Human-controlled mast-free sailing platforms allow long distance travelling in a wide range of weather conditions [2]. The task is to make the sailing platform completely autonomous.

To implement the mast-free autonomous sailing platform, the specific control system is required. This control system shall stabilize the platform rectilinear uniform motion under the small wind velocity and direction fluctuations conditions and under weak wave conditions. These are the control system minimal requirements. The strong waves control is the different task and is out of current paper consideration.

Kite towed object system model is complex and multiparametric [3]. Hence, complex regulators are required to control that system. This paper describes an attempt to design a simple control system. Proposed control system is based on human manual control experience.

Control system consists of few independent regulators. Each regulator should have only one input parameter, as in that case the system can be manually tuned without complicated stability calculations.

2 Mast-Free Sailing Platform Model

Mast-free sailing platform consists of two parts: a kite and a board. The kite and the board are connected with inextensible mechanical link that is referred to as line. Board is partially drowned. Kite glides above the water surface. It can easily rotate around the board connected to kite. Herewith the distance between the kite and the board is constant in any kite rotation angle condition. Kite-board system has some translator motion velocity with relation to water surface. This system is similar to the system, proposed in [4], but the towed object is not the hydrofoil; in the present paper the loaded kite-board is used. The human controlled system example is presented on Fig. 1. If human is replaced by the automatics, the forces fulcrums shifts to a single point.

2.1 General Kite Model Description

For mechanic tasks kite can be presented by model that consists of forces and torque concentrated in single point. The model feature, as in [5], is the replacement of complex kite air flow Navier Stokes equations solution with constant coefficients. In the model of [5] the system has no translatory movement, but the kite moves within the wind window.

Fig. 1 Human controlled board forces and axes

The proposed model considers system translatory movement, but it neglects the kite movement within the wind window. This assumption is acceptable for the steady states simulations, when kite moves within wind window slowly.

2.2 *General Board Model Description*

A board is water surface planing object towed by kite. Hence, board mechanic model should consider water planing hydrodynamic force, gravity force and kite tow force. The forces fulcrums do not match so each force causes a torque.

Water planing hydrodynamic force is directed orthogonally to board surface, as shown on Fig. 1. Its magnitude can be accurately calculated by water flow equations with considering board geometry. However, for mechanical tasks, the force can be approximately calculated by square of water surface velocity and board wetted surface area [6]. The force fulcrums is a center of board wetted surface geometry figure. On the Fig. 1 the wetted surface shown as figure shaded with red lines.

$$F_{hd} = C_{hd} \frac{\rho_w V_d^2}{2} S_d \qquad (1)$$

where C_{hd} is hydrodynamic coefficient, ρ_w is water density, V_d is water surface board velocity and S_d is board wetted surface area.

Table 1 Kite model assumptions

Assumption	Grounds/comments
Kite is absolutely rigid	Real kite wing is semi-rigid. However, in fly time, relative changes in kite linear dimensions are units of percent. So kite flat area, determined by the dimensions, changes in units of percent also. Aerodynamic force is proportional to flat area, so its error is not more than units of percent. Small linear dimensions change is provided by kite producer because kite is controllable only when its geometry is maintained
Kite is joined to board by single line	Real kite has two power lines and two control lines. However, control lines do not participate in force transmission. Two real power lines can be replaced by single virtual line. This virtual line begins on real fulcrum on board and finishes in kite fulcrum
Line is joined to kite in aerodynamic force fulcrum	Kite aerodynamic force fulcrum also is a virtual point. This force is distributed on its surface. However, kite is constructed to provide match of equivalent aerodynamic force fulcrum and the virtual line in all states except the rotations around the line state
Kite is controlled by two kite angles set: kite open angle and kite yaw	In real kite there is no direct controller of yaw. The control is realized by alteration of two control lines' lengths. Control lines instantly and directly affect kite open angle and indirectly set kite yaw with a small time constant
Kite aerodynamic force is function of aerodynamic coefficient and kite flat area. Aerodynamic coefficient and flat area are functions of angle of attack	In the model air flow calculations are avoided. The coefficients are found for all incoming air flow directions
Gravity center is situated on kite fulcrum	Real kite is constructed to avoid extra torques caused by mismatch of gravity fulcrum and kite aerodynamic fulcrum. However, automation devices placed on kite can shift gravity fulcrum. In this case the additional torques should be considered. Especially, if kite with automatic device weight is much more than kite own weight

2.3 *Mast-Free Platform Model Assumptions*

To use simulation results correctly it is necessary to know the assumptions that are accepted in the model. Presented model considers nonzero system movement states only. Assumptions for each element of the model are presented in the Tables 1, 2 and 3 with short comments.

Table 2 Kite line model

Assumption	Grounds/comments
Line is inextensible	The relative tensile of 1.4 mm diameter "Dyneema" power line usually used in kite design is less than 5 % [7]
Line weight can be neglected	Running weight of 1.4 mm diameter "Dyneema" line is 1.2 grams per meter [7]
Line is always stretched	The model is valid for line stretched states. If line is not stretched then kite is uncontrollable. In this state, it is possible to simulate its fall only. The paper doesn't consider that state

Table 3 Board model

Assumption	Grounds/comments
Board is absolutely rigid	It is true for surfboards or wake boards. Most kiteboards are flexible
Hydrodynamic force is concentrated in single point of board wetted surface. In simulations the force is replaced by force on board gravity center and corresponding torque	In reality, the force is irregularly distributed on wetted surface. For simple geometry board it is fore of board wetted surface. For complicated geometry board, concentrated hydrodynamic force fulcrum moves with respect to angles of attack
Water planning hydrodynamic force is directed orthogonally to board plane	In reality, in the board plane the extra hydrodynamic force component exists

2.4 Forces and Torques in Kite Axes System

Kite axes system consists of three orthogonal axes. The first one matches with kite line, second one is directed along a kite central foil and third one is orthogonal to the central foil plane. Axes system origin matches with kite fulcrum. The axes system is presented on Fig. 1.

Kite simulation input parameters (Fig. 2)

- $F_T^b = \{ F_{Tx}^b; \quad F_{Ty}^b; \quad F_{Tz}^b \}$—line tension force vector in board coordinate system,
- $V_w^{ws} = \{ V_{wx}^{ws}; \quad V_{wy}^{ws}; \quad 0 \}$—wind velocity vector in water surface coordinate system,
- $V_d^{ws} = \{ V_{dx}^{ws}; \quad V_{dy}^{ws}; \quad 0 \}$—board velocity vector in water surface coordinate system,
- $A^{b-k} = \begin{bmatrix} A_{xx}^{b-k} & A_{xy}^{b-k} & A_{xz}^{b-k} \\ A_{yx}^{b-k} & A_{yy}^{b-k} & A_{yz}^{b-k} \\ A_{zx}^{b-k} & A_{zy}^{b-k} & A_{zz}^{b-k} \end{bmatrix}$—board to kite coordinate systems conversion operator,

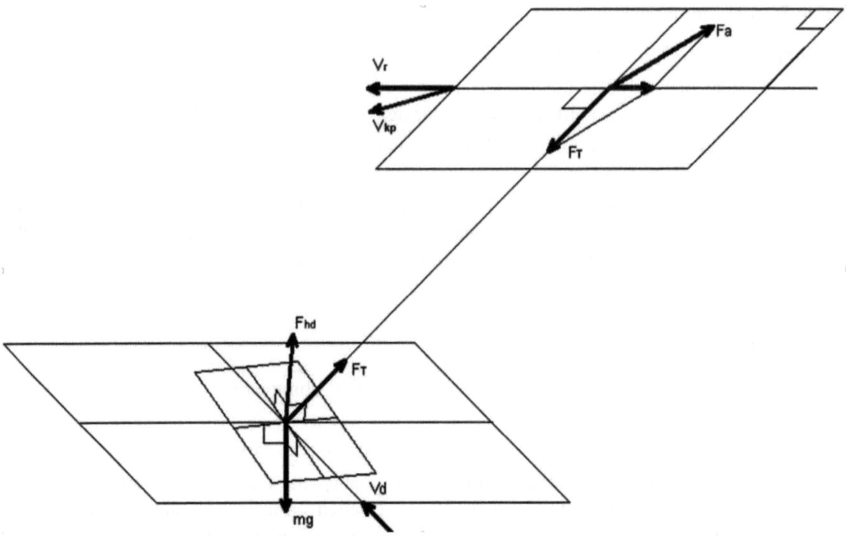

Fig. 2 Mast-free sailing platform model

- $A^{ws-k} = \begin{bmatrix} A_{xx}^{ws-k} & A_{xy}^{ws-k} & A_{xz}^{ws-k} \\ A_{yx}^{ws-k} & A_{yy}^{ws-k} & A_{yz}^{ws-k} \\ A_{zx}^{ws-k} & A_{zy}^{ws-k} & A_{zz}^{ws-k} \end{bmatrix}$—water surface to kite coordinate systems conversion operator.

Kite simulation internal parameters

- $F_T^k = \{ F_{Tx}^k; \quad F_{Ty}^k; \quad F_{Tz}^k \}$—line tension force vector in kite coordinate system,
- $F_a^k = \{ F_{ax}^k; \quad F_{ay}^k; \quad F_{az}^k \}$—aerodynamic force vector in kite coordinate system,
- $V_r^k = \{ V_{rx}^k; \quad V_{ry}^k; \quad V_{rz}^k \}$—kite rotational motion with respect to board velocity vector in kite coordinate system,
- $a^k = \left\{ a_x^k; \ a_y^{kk}; \ a_z^k \right\}$—kite acceleration vector.

2.5 Kite Motion Parameters Calculation Algorithm in Kite Coordinate System

To calculate a time step of kite simulation it is necessary to use kite model input parameters from board model that were calculated on previous time step and use it in kite equation in correct order, presented below and illustrated in Fig. 2.

1. Line tension force vector is converted to kite coordinate system (initially line tension force was board axis vector)

$$F_T^k = A^{b-k} \cdot F_T^b \tag{2}$$

2. The mast-free platform velocity vector relative air flow V_{kp}^{ws} is obtained by sum of board velocity vector V_d^{ws} and wind velocity vector V_w^{ws}. This operation is performed in water surface coordinate system

$$V_{kp}^{ws} = V_w^{ws} + V_d^{ws} \tag{3}$$

3. The mast-free platform velocity vector V_{kp}^{ws} is converted to kite axes

$$V_{kp}^{k} = A^{ws-k} \cdot V_{kp}^{ws} \tag{4}$$

4. Incoming air flow vector in kite coordinate system V^k is obtained by sum of the mast-free platform velocity vector V_{kp}^k and board relative kite rotational motion velocity vector V_r^k (obtained from previous simulation step)

$$V^k = V_{kp}^k + V_r^k \tag{5}$$

5. Incoming air flow vector direction angle is calculated by current kite open angle α and incoming air flow vector direction

$$\gamma = \alpha + arctg\left(\frac{V_y^k}{V_x^k}\right) \tag{6}$$

6. Aerodynamic force coefficients components and flat area are obtained by incoming air flow vector direction angle (kite angle of attack) $C^{xy} = f(\gamma)$. The coefficient and flat area values is the table function. The values are obtained by the measurements. The table structure is presented in Table 4.

7. Aerodynamic force is calculated by incoming air flow magnitude $\left|V_{xy}^k\right| = \sqrt{\left(V_x^k\right)^2 + \left(V_y^k\right)^2}$ and ρ_{air}—density of air

$$\overline{F_a} = \left\{F_a^x, F_a^y, F_a^z\right\} = \left\{C^x(\gamma), C^y(\gamma), 0\right\} \cdot \left|\overline{V_{xy}^k}\right|^2 S_{fa}(\gamma) \cdot \rho_{air} \tag{7}$$

Table 4 Aerodynamic force coefficients components and flat area table

γ(degrees)	Flat area	C^{xy}
− 30	S_{fa}	$\{C^x, C^y\}$
…	…	…
30	S_{fa}	$\{C^x, C^y\}$

2.6 Forces and Torques in Board Coordinate System

Board simulation input parameters

- $F_T^b = \{F_{Tx}^b; \quad F_{Ty}^b; \quad F_{Tz}^b\}$—kite tow force vector in board coordinate system is obtained from kite simulation.
- $M_T^b = \{M_{Tx}^b; \quad M_{Ty}^b; \quad M_{Tz}^b\}$—kite tow torque vector in board coordinate system.

Board simulation internal parameters

- $F_{hd}^b = \{F_{hdx}^b; \quad F_{hdy}^b; \quad F_{hdz}^b\}$—hydrodynamic force vector in board coordinate system.
- $m^b g^b = \{m^b g_x^b; \quad m^b g_y^b; \quad m^b g_z^b\}$—gravity force vector in board coordinate system.
- $a^b = \{a_x^b; \quad a_y^b; \quad a_z^b\}$—board acceleration vector in board coordinate system.
- $M_{hd}^b = \{M_{hdx}^b; \quad M_{hdy}^b; \quad M_{hdz}^b\}$—hydrodynamic torque vector in board coordinate system.
- $M_{mg}^b = \{M_{mgx}^b; \quad M_{mgy}^b; \quad M_{mgz}^b\}$—gravity torque vector in board coordinate system.
- $\varepsilon^b = \{\varepsilon_x^b; \quad \varepsilon_y^b; \quad \varepsilon_z^b\}$—board angular acceleration vector in board coordinate system.
- $r^b = \{r_x^b; \quad r_y^b; \quad r_z^b\}$—force radius vector in board coordinate system.
- $J^b = \{J_x^b; \quad J_y^b; \quad J_z^b\}$—rotational inertia vector in board coordinate system.

The board translatory motion equation

$$F_T^b + F_{hd}^b + m^b g^b = m^b a^b \tag{8}$$

The board rotational motion equation

$$M_T^b + M_{hd}^b + M_{mg}^b = J^b \varepsilon^b \tag{9}$$

The systems can be simplified if hydrodynamic force vector is orthogonal to board plane (as shown in Fig. 3).

- $F_{hd}^b = \{0; \quad 0; \quad F_{hdz}^b\}$.
- $M_{hd}^b = \{M_{hdx}^b; \quad M_{hdy}^b; \quad 0\}$.

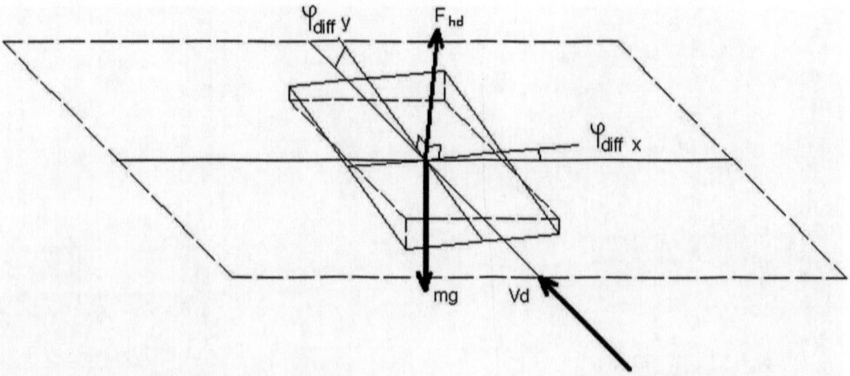

Fig. 3 Board forces model

So, the board torque can be represented as (as shown in Fig. 4):

$$\begin{cases} r^b_{Tx}F^b_{Tz} + r^b_{hdx}F^b_{hdz} + r^b_{mgx}mg^b_z = J^b_x\varepsilon^b_x \\ r^b_{Ty}F^b_{Tz} + r^b_{hdy}F^b_{hdz} + r^b_{mgy}mg^b_z = J^b_y\varepsilon^b_y \end{cases} \tag{10}$$

If the board gravity center matches with board fulcrum one can avoid the correspondent torque use:

$$\begin{cases} r^b_{Tx}F^b_{Tz} + r^b_{hdx}F^b_{hdz} = J^b_x\varepsilon^b_x \\ r^b_{Ty}F^b_{Tz} + r^b_{hdy}F^b_{hdz} = J^b_y\varepsilon^b_y \end{cases} \tag{11}$$

If board type is surf board, its horizontal rotation axis matches with the board aft fin. In the case of horizontal rotation, equation has different radius vectors than radius vectors in first equations

$$r^b_{Tz}F^b_{Ty} + r^b_{hdz}F^b_{hdy} + r^b_{mgz}mg^b_y = J^b_z\varepsilon^b_z \tag{12}$$

If hydrodynamic force vector is orthogonal to board plane, its horizontal projection is zero $F^b_{hdy} = 0$ and equation becomes simpler:

$$r^b_{Tz}F^b_{Ty} + r^b_{mgz}mg^b_y = J^b_z\varepsilon^b_z \tag{13}$$

For stabilization simulation, board rotational inertia can be approximately calculated as

$$J^b_x \cong m_b\frac{W^2_b}{12} \tag{14}$$

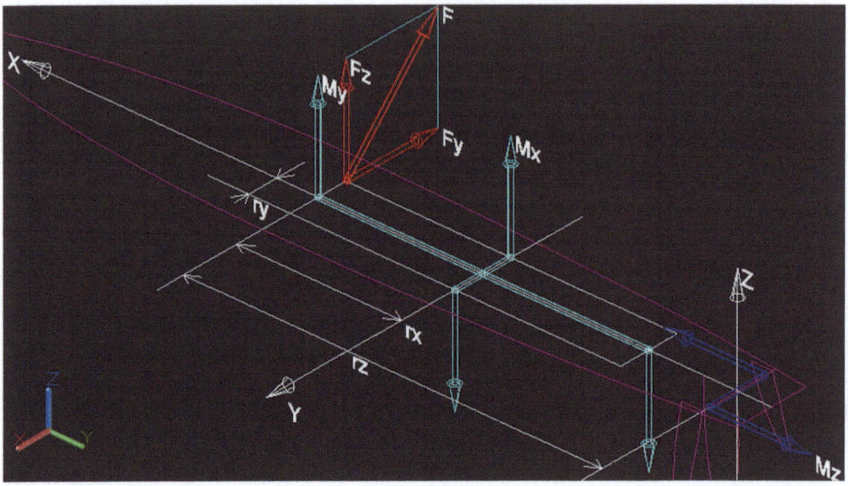

Fig. 4 Radius vectors and torques of board force

$$J_y^b \cong m_b \frac{L_b^2}{12} \tag{15}$$

$$J_z^b \cong m_b \left(\frac{L_b^2}{12} + \frac{W_b^2}{3} \right) \tag{16}$$

where L_b is board length, W_b is board width.

3 Mast-Free Sailing Platform Practical Stabilization Method Based on Kite Manual Control Experience

3.1 General Idea of Independent Parameters Control

Presented above model is a complicated dynamic system. Design stabilization control for the system is not trivial task. However, on practice the kite-board system is easily stabilized by a human. Nevertheless, human uses another simplified model.

According to authors' experience, manual stabilization is based on independent stabilizations of each controlled parameter. Human controls kite tow power and kite position separately. It is possible because kite power regulation and kite position regulation have different time constants. Kite tow force changes almost instantly after control action. It is provided by kite design. Hence, kite moves slowly enough with respect to board with the time constants conditioned by line length (from 25 to 30 meters). [8].

Therefore, human regulates all kite-board system parameters with different priorities, fast alternating parameters first. Human controls kite tow power constantly. If power is acceptable, human controls kite position. If power is different from the acceptable, then human corrects kite power. In this method, kite power regulation process doesn't overlap kite position regulation.

Manual board control is implemented at the similar way. Human moves kite line fulcrum along the board. It causes additional torque. The torque changes board course. Human continuously controls board hydrodynamic force also. To implement that control, human moves kite line fulcrum from the center to the side of the board. This action changes board angle of attack that directly influences hydrodynamic force magnitude. Hydrodynamic force regulation speed is higher than board course regulation speed. Board width is about 0.4 m so hydrodynamic force regulation takes a split second, by authors' experience. Course correction takes seconds because it is conditioned by board rotation radius measured in tens of meters on nominal speed. So, hydrodynamic force and board course can be regulated independently. [8].

3.2 Implementation of Independent Parameters Control in Automatic System

Automatic control system for mast-free sailing platform could be implemented by the same algorithm as human manual control described above. Multiple regulators could control the whole system independently if each regulator acts on the different frequencies [9, 10]. The slow regulator is not capable to perform significant control action affecting the fast alternating parameter. The fast regulator could approximately consider the slow alternating parameter as constant or nonalterating at all. Therefore it doesn't affect the alteration of slow alternating parameter. Therefore, the independency of control actions is performed if the regulators act on the significantly different frequencies.

Mast-free sailing platform automatic regulation system includes four independent regulators. Each regulator is simple PI regulator with only one controlled parameter and only one control action. Other possible control actions are neglected in the proposed control system.

The first regulator presented on the Fig. 5 is used for kite tow force stabilization. Tow force is measured by tension sensor mounted on board. Control action is kite open angle (angle between kite line and kite air foil axis, or α). Regulation time delay is practically zero.

The second regulator presented on Fig. 6 is used for kite position stabilization. Regulated kite position angle is calculated by three axes accelerometer measured parameters. Control action of the regulator is indirect. Regulation delay is measured in seconds. Controlled parameter is filtered by frequency filter with cut off level corresponding fluctuations period measured in seconds. Therefore, the control will be independent from kite tow force stabilization.

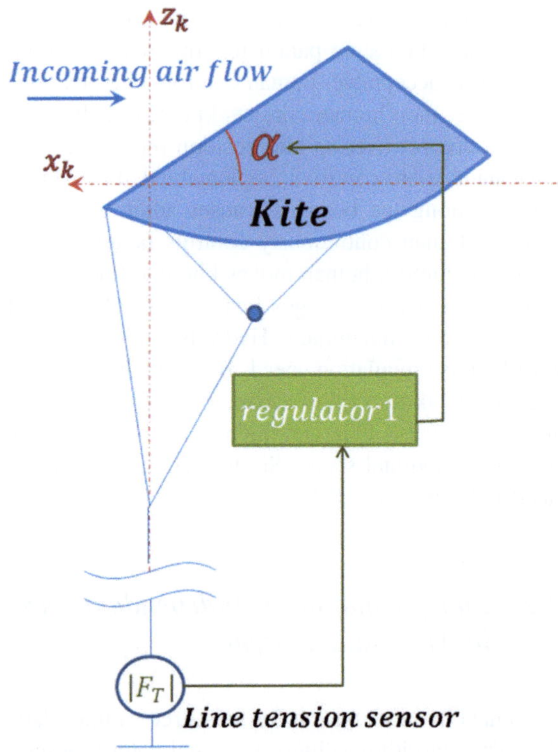

Fig. 5 Kite tow force regulator block diagram

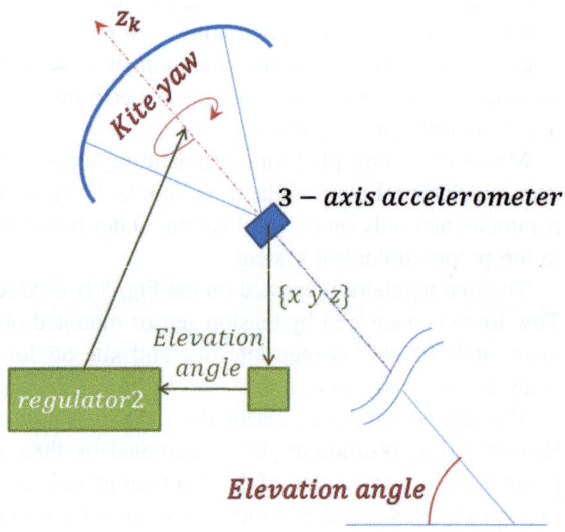

Fig. 6 Kite elevation regulator block diagram

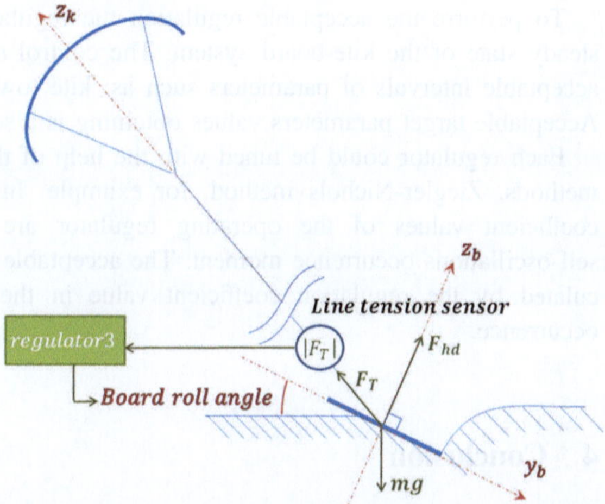

Fig. 7 Board roll regulator block diagram

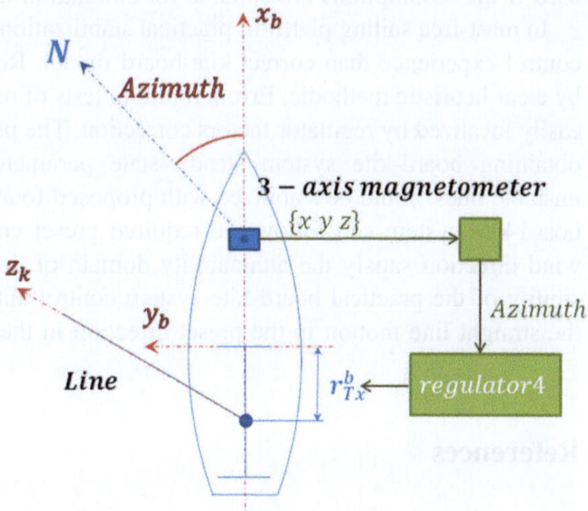

Fig. 8 Board course regulator block diagram

The third regulator presented on Fig. 7 is used for board drowning stabilization. Control parameter is kite line tow force. It should compensate board hydrodynamic force correspondent component. Control action is line fulcrum transverse shift. Time delay for the regulator should be higher than kite position regulator to implement the independency of regulations. The regulator should set correct board attack angle when kite position is set.

The fourth regulator presented on Fig. 8 is used for board course stabilization. Control parameter is course obtained from magnetic sensor. Control action is line fulcrum lengthwise shift. Time delay for the regulator is the highest, and is about tens of seconds.

To perform the acceptable regulation the regulators should be tuned for the steady state of the kite-board system. The control system requires target values' acceptable intervals of parameters such as: kite tow, kite position, board course. Acceptable target parameters values obtaining is a separate task.

Each regulator could be tuned with the help of the relatively simple empirical methods, Ziegler-Nichols method, for example. In the methods, the regulation coefficient values of the operating regulator are gradually increased to the self-oscillations occurrence moment. The acceptable regulator coefficients are calculated by the regulation coefficient value in the moment of self-oscillations occurrence.

4 Conclusion

Mast-free sailing platform simulation is a complicated mechanical task. However, it can be solved if assumptions are pointed out. Solution results can be practically used if the assumptions are suitable for simulation target.

In mast-free sailing platform practical stabilization task it is easier to use manual control experience than correct kite-board model. Regulation system can be tuned by clear heuristic methodic. Errors found in tests of regulation system tuning can be easily localized by regulator factors correction. The proposed model analysis allows obtaining board-kite system steady state parameters. The steady states, even unstable ones, could be stabilized with proposed four regulators use. The stabilized board-kite system can follow the required preset course if the preset course and wind direction satisfy the attainability domain of the model. Hence, there is possibility of the practical board-kite system control automatics realization providing the straight line motion in the preset direction in the weak waves condition.

References

1. The Official Website of the International Kiteboarding Association: http://internationalkiteboarding.org/
2. Guinness World Records: http://www.guinnessworldrecords.com/world-records/longest-journey-kite-surfing-(male)
3. Loyd ML (1980) Crosswind kite power (for large-scale wind power production). J Energy 4 (3):106–111
4. Nedeleg B et al (2015) 6 DOF Simulations for Stability Analysis of a Hydrofoil Towed by Kite. In: 5th high performance Yacht design conference. Auckland, 10–12 March 2015
5. Leloup R et al (2014) Estimation of the Lift-to-Drag ratio using the Lifting line method: application to a leading edge inflatable kite. In: Robotic airborne wind energy. Springer, Heidelberg, pp. 339–359
6. Bertorello C, Oliviero L (2008) Surf hydrodynamics. In: Hiper 08 international conference 17–19 Sept 2008, Napoli, Italia, pp. 281–290
7. Dyneema Official Site: http://www.dsm.com/products/dyneema/en_GB/home.html

8. Cabrinha Official Site: http://pr.cabrinhakites.com
9. Zaicev P, Formalskii A (2008) Independent longitudinal movement of the wing: mathematical modeling, control synthesis. Russian Academy of Sciences, theory and control systems
10. Ivanov P (1996) Design, creation and testing of the wing. Moscow

Part II
Mission Planning, Localization and Obstacle Avoidance

AIS-Enabled Collision Avoidance Strategies for Autonomous Sailboats

José C. Alves and Nuno A. Cruz

Abstract Autonomous sailing boats are a valuable resource for ocean monitoring, sampling and surveillance due to their intrinsic very low energy requirements and extreme capability for long term navigation. However, the extended presence in the ocean is vulnerable to a multitude of hazards that may compromise the success of a mission. Although the conventional manned sea crafts must comply to the international rules of the road (or COLREGS), unmanned (small and slow) robotic boats should actively anticipate close encounters with (large and fast) ships that in practical situations may not be able to avoid collisions with small vessels. This paper proposes a set of simple rules for implementing a defensive and conservative collision avoidance strategy for autonomous sailing boats, based solely on the AIS data received from large ships.

1 Introduction

Unmanned autonomous sailing boats are a proven technology for ocean monitoring, sampling and surveillance due to their intrinsic very low energy requirements, good navigation ability and extreme potential for long navigation journeys [1]. However, maintaining an operation in the sea surface for long periods of time is susceptible to various risks of failure due to external threats. Environmental conditions can now be predicted for a few days window with high degrees of confidence, thus allowing re-plan a mission in advance to avoid potential dangerous wind and sea states. However, the risk of collision with other vessels or floating debris represents a significant menace especially when operating in regions of intense marine traffic, like the English Channel or the west coast of Portugal and Spain.

The detection of small boats, floating fishing nets or partially submerged wrecks (e.g. lost containers) requires onboard active devices like radar, image or acoustic

J.C. Alves (✉) · N.A. Cruz (✉)
FEUP/INESC TEC, Rua Dr Roberto Frias, 4200-465 Porto, Portugal
e-mail: jca@fe.up.pt

N.A. Cruz
e-mail: nacruz@fe.up.pt

A. Friebe and F. Haug (eds.), *Robotic Sailing 2015*,
DOI 10.1007/978-3-319-23335-2_6

sensors and the effective detection range depends significantly on the size of the boat (more precisely the height of the sensors), the state of the sea and, for underwater imaging, the transparency of the water. Although the avoidance of collisions with such small objects may be effectively done with last minute maneuvers executed at a short distance of the object (in a practical range of tens of meters), the situation with large ships must be handled in a very different manner in order to ensure a comfortable distance of safety, that depends on the size of the meeting ships and hence on their relative maneuverability. This is a general principle clearly stated in the COLREGS [2] and adopted by captains of real ships: for example, if 1 m may be a safe minimum distance between two small sailing dinghies, a 100 m separation between two large tankers sailing at cruising speed is certainly considered a severe near miss. Thus, an effective mechanism for avoiding collisions with large ships[1] must be able to guarantee the appropriate means for executing a convenient and timely maneuver to improve the safety margin.

Although all ships must obey to the international collision avoidance regulations (COLREGS), when ships meet in a potential collision route the decision to maneuver for avoiding the collision may be dictated by other factors than the simple right of way stated by the rules. For example, both ships must be aware of the speed, course and approximate position of each other, and the relative maneuverability of the meeting ships must be taken into account. Considering situations involving the encounter between a small (and slow) robotic vessel (ASV) and a large (and fast) ship, some realistic assumptions need to be taken into consideration:

- Large ships navigate most of the time under autopilot and human surveillance. A small and possibly intermittent radar echo may be ignored if it is not visually confirmed by the crew.
- A large ship will only have clear visual contact with a small ASV within a few hundred meters range (by night the navigation lights may improve the visibility). This may be too close for effectively maneuvering to avoid a collision.
- Small ASVs, and specially robotic sailing vessels, may not be able to carry or generate enough energy to power a AIS (Automatic Identification System) transponder, so they will not be detected at large distances from other vessels.

If the robotic vessel is a sailing boat, several additional constraints must be considered for implementing collision avoidance maneuvers, in spite of the default rule stated in COLREGS giving the right-of-way to sailing ships. A sailing boat cannot naturally navigate outside of the feasible wind angles, cannot increase its speed (usually sailing boats always seek its maximum speed) and in certain situations a sailing boat may also not be able to reduce its speed, as for example in a dead downwind course with certain types of rigs.

Presently the situational picture of the surrounding (large) vessels can be easily done by decoding the AIS (Automatic Identification System) messages transmitted in the maritime VHF band. Although the transmission of AIS data is not yet compulsory for all ships, presently all commercial ships and fishing vessels longer than 16 m must

[1]The term *large* must be understood in relation to the size of the autonomous boat.

broadcast their navigation status using the AIS. With the continuous reduction of the cost of transponders, many pleasure vessels also opt to use AIS as an additional safety asset and in the next years it is expected that more and more boats will be equipped with AIS. Even though to transmit AIS data it is required a certified equipment, an operating license and a unique ship identification number (MMSI—Maritime Mobile Service Identity), the reception of AIS data only needs a dedicated AIS receiver or a radio receiver for the AIS radio frequencies and software to decode, extract and interpret the AIS data.

To minimize the risk of collision with other ships, a (small) robotic sailing craft can adopt a defensive behavior to avoid possible near miss situations and anticipate maneuvers for reducing the risk of collision. In this paper we propose a defensive navigation strategy for collision avoidance, based on the AIS received data or by any other means capable of providing quasi-real-time the position, course and speed of other vessels. Instead to comply to the COLREGS, which in most situations gives the right-of-way to sailing boats, our strategy is to actively avoid any known AIS target, regardless of the type of vessel and its possible change of behavior to give the way.

The rest of the paper is organized as follows. Section 2 reviews some efforts towards the automation of the application of COLREGS for autonomous ships. Section 3 includes a brief description of the AIS system and references to software-defined radio projects implementing AIS receivers. The proposed strategies for defensive collision avoidance are discussed in Sect. 4 and the paper is closed in Sect. 5 with the concluding remarks and proposals for future directions.

2 Automating COLREGS for Robotic Surface Vessels

It is clear that the current technology status of various perception systems allow ASVs to build a clear situational picture of its surroundings and provide to the computing system driving it the necessary information for avoiding collisions with other ships [3]. To be accepted in environments shared with conventional manned crafts they cannot be considered a potential danger for people or property. Many works have been proposing various techniques for implementing the rules stated in COLREGS without direct human intervention. However, the interpretation of these rules is not strict and in many cases the common sense of captains is crucial to assess practical situations and avoid collisions.

As discussed in [4], the COLREGS were designed for being used by humans and contain ambiguities that must be interpreted with a certain degree of flexibility. In that work the authors employ a behavior-based control architecture to coordinate behaviors that lead to a flexible application of the rules, using an interval programming procedure for multi-objective optimization.

Other techniques exploit Genetic Algorithms (GAs) to generate near optimal COLREGs-compliant escape maneuvers [5], fuzzy logic theory [6], neural networks [7] or line-of-sight navigation with course biasing to deviate from detected obstacles [8].

A recent work [9] uses the concept of Velocity Obstacles to implement a real-time navigation planning system for roughly equivalent size and power motorized vessels, considering uncertainty in the speed and course of the meeting boats. As in several other works, this is focused on scenarios where the ASV navigates close to other vessels and perceive its surrounding environment with onboard sensors (in this case a stereo vision system).

Although many work has been done on adapting the human-based COLREGS to automatic unmanned surface vessels, the application of the rules by the boats involved in a potential collision route assumes that each vessel is able to observe the other and give the right of way when obliged by the rules. However, when the sizes, and consequently maneuverability, of two approaching boats are very different, the safest rule to apply for the smaller boat is to go away timely, regardless of what is written in COLREGS.

3 The Automatic Identification System—AIS

AIS is a ship location and tracking system mandatory to commercial ships above 300 tons under international navigation, all passenger ships and fishing ships above 16 m in length [10]. As the prices of the AIS equipments drop, this technology is being more and more adopted by smaller and pleasure ships. The system uses two radio channels in the maritime VHF band and transmits periodically various short messages reporting ship's data and its navigation status. The most important AIS message broadcasts the ship identification, navigational status, position, course and speed, which is sufficient for evaluating the potential risk of collision.

Although there are in the market various AIS receivers that are becoming cheaper, smaller and less power hungry, one alternative solution is the implementation of a fully software-defined AIS receiver in the sailboat's onboard computer. The recent launch of the low cost RTL2832/R820T/E4000-based universal radio receiver USB dongles has enabled the development of various software-defined radio projects capable of running in embedded low power computers like the RaspberryPi or the BeagleBone, including a complete AIS receiver and decoder chain (eg. the GNU AIS project—gnuais.sourceforge.net). AIS data is becoming affordable to small and power constrained robotic sailing boats and is thus an important resource for mitigating the risk of collision with medium to large ships.

4 Strategies for Defensive Collision Avoidance

4.1 Why Defensive Collision Avoidance?

In this work we consider situations arising from a potential risk of collision when a small sailing ASV encounters a large ship reporting AIS data. We assume the sailing ASV is receiving the AIS data (but not transmitting), or it has any other perception system capable of providing in real-time the location, speed and course information of the neighbor vessels. Although the location of the GPS antenna in the reporting ship is part of the AIS messages, we consider only the reported geographic position and the boat size is not relevant for the large minimum safe distance.

Instead of embedding in the navigation control system the mechanisms to apply the rules of priority established by COLREGS, we adopt an early defensive behavior to reduce the risk of collision and assume that the ASV will *never* have right-of-way with respect to the other vessels. This can be justified with the following arguments:

- To make use of a eventual right-of-way, a ASV must be clearly identified by the other vessel. In practical situations it may be difficult for a large ship to establish visual contact at a safe distance to allow the implementation of give-way maneuvers. On the other hand, a ASV will perceive (using the AIS data) the other ships much sooner and will have enough time to alter its navigation to avoid the encounter.
- Even if the crew of a large ship identifies a small sailboat in route of collision, with evidences of being unmanned and harmless to the ship, it is very unlikely they will change course to avoid the collision.

Additionally, a small sailing ASV is prone to the wind turbulence caused by the hull of a large ship. This wind shadow can extend leeward for a distance significantly longer than the length of the hull, affecting the sailing ability of the ASV to perform a last minute avoidance maneuver. It is thus desirable that the ASV should actively maneuver to maintain a safe distance from any moving ship significantly larger than the minimum to comfortably avoid a physical contact.

To keep clear of the other ships, a robotic sailing boat can, in some courses, reduce its speed by luffing the sails while keeping its course, or temporarily deviate from the risky route. Although a sailing robot can only navigate within the feasible points of sailing, we consider that if a escape route requires a temporary upwind course, that will be done by tacking along a narrow corridor with a few tens of meters wide. Also, the collision avoidance maneuver should not increase the risk of collision in the event of the other ship decided to maneuver to deviate, according to the rules stated in COLREGS.

4.2 Assumptions and Definitions

In this work we consider a scenario where an ASV A and a large ship S navigate along different but fixed courses and move with constant speeds. To keep the computational effort low and affordable to low performance microcontrollers, our approach is based on the continuous computation of the closest point of approach for both vessels (X_{CPA}, Y_{CPA}), the time to reach that point ($TCPA$) and the minimum distance between them ($MDCPA$), assuming that the wind speed, wind direction and sea state, which dictates the effective speed of the sailing boat along its course, will stay constant during the time interval required for the collision avoidance maneuver. For now we consider both vessels navigate through a straight line, and later extend to the case when a sailboat navigates upwind tacking periodically within a narrow corridor. Other variables not considered explicitly in this simple formulation (sea currents, the boat's kinematics and dynamics) are indirectly considered as we assume both ships are sailing in a straight line course with constant speed.

If the current positions of the vessels are (X_a, Y_a) and (X_s, Y_s), and v_{x*} and v_{y*} represent the X, Y components of the velocities of the two vessels, the future distance between them, $d_{as}(t)$ is given by:

$$d_{as}(t) = \sqrt{[(X_a + v_{xa} \cdot t) - (X_s + v_{xs} \cdot t)]^2 + [(Y_a + v_{ya} \cdot t) - (Y_s + v_{ys} \cdot t)]^2} \quad (1)$$

The time to the closest point of approach ($TCPA$) and the minimum distance of approach ($MDCPA$) can then be calculated by minimizing function $d_{as}(t)$. Defining the difference of velocities and positions along X and Y as $d_{vx} = v_{xa} - v_{xs}$, $d_{vy} = v_{ya} - v_{ys}$, $d_x = X_a - X_s$ and $d_y = Y_a - Y_s$ we obtain:

$$TCPA = -(d_{vx} \cdot d_x + d_{vy} \cdot d_y)/(d_{vx}^2 + d_{vy}^2) \quad (2)$$

$$MDCPA = \sqrt{(d_{vx}^2 + d_{vy}^2) \cdot TCPA^2 + (d_{vx} \cdot d_x + d_{vy} \cdot d_y) \cdot TCPA + (d_x^2 + d_y^2)} \quad (3)$$

$$X_{a_{CPA}} = X_a + v_{xa} \cdot TCPA \quad (4)$$

$$Y_{a_{CPA}} = Y_a + v_{ya} \cdot TCPA \quad (5)$$

$$X_{s_{CPA}} = X_s + v_{xs} \cdot TCPA \quad (6)$$

$$Y_{s_{CPA}} = Y_s + v_{ys} \cdot TCPA \quad (7)$$

If the courses are not parallel, a positive $TCPA$ means that the minimum distance of approach $MDCPA$ will happen at time $TCPA$ in the future. If $TCPA$ is negative then the positions of the two vessels are monotonically diverging and there is no risk of collision.

When the sailboat navigates an upwind route, tacking periodically, the exact calculation of the previous parameters is far more complex as it depends on the geographic positions where the sailboat will effectively tack. When a robotic sailboat sail upwind for long distances in the same tack, this can be treated as if it is

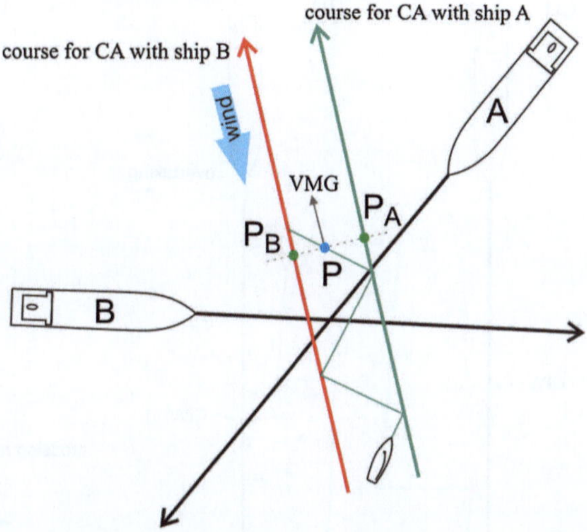

Fig. 1 Approximating the course upwind of the sailboat ASV to the limits of the upwind corridor

following a straight line. In the case an upwind navigation is constrained to a narrow corridor (tens to few hundreds of meters) we apply a conservative but simple rule: approximate the course of the sailboat to the straight line defining the side of the corridor closer to the approaching ship, and the sailboat speed equal to the average VMG observed along that route in the present conditions. As illustrated in Fig. 1, when the sailboat is in position P it is considered in position P_A (P_B) for analysing the *CPA* with ship A (ship B).

4.3 Collision Avoidance Maneuvers

To reduce the risk of a collision with a ship, the robotic sailboat performs two basic maneuvers: if possible and advantageous to increase *MDCPA*, keep the course and reduce speed until *MDCPA* increases above the minimum acceptable; if the speed cannot be reduced while keeping the course (for example in a dead downwind course) or if slowing down will further reduce *MDCPA*, then set an escape course COG_e perpendicular to the ship's course, along a direction that will not cross it. While these behaviors naturally will not guarantee the avoidance of a collision, neither prove that a collision can not be avoided, they provide a very simple mechanism to mitigate practical situations when tho very dissimilar ships meet in the ocean. Besides, in this work we only address the cases when a single ship is in potential risk of collision with a ASV, what may be too restrictive under situations of dense marine traffic.

As in COLREGS, we distinguish three different situations with respect to the relative courses of the approaching ships: heading on, overtaking and crossing routes. In the first case (heading on or opposite courses, Fig. 2a), it is clear that reducing the speed of the sailboat will not contribute significantly to increase *MDCPA*. The same

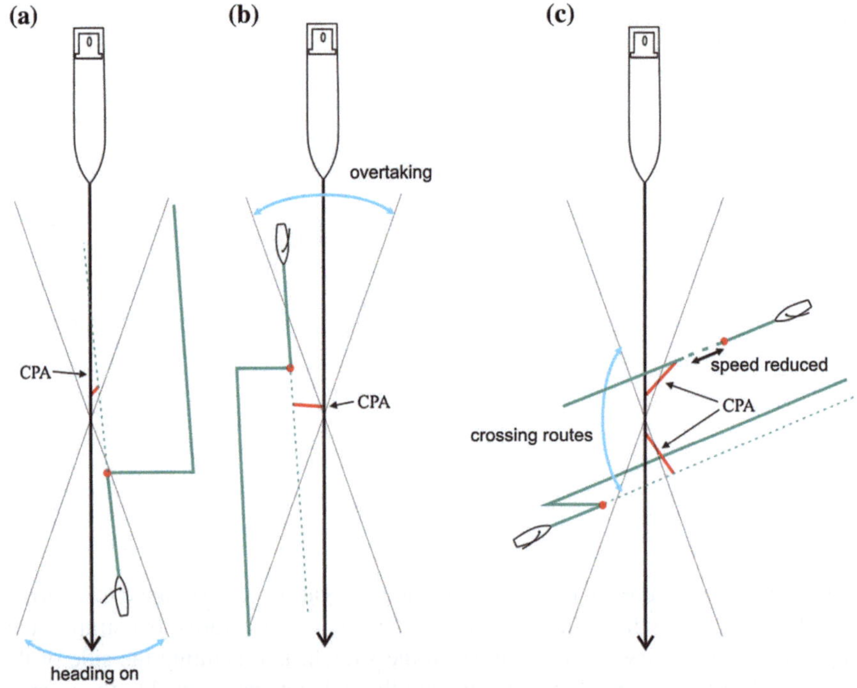

Fig. 2 The basic collision avoidance maneuvers

happens in the second situation (similar courses, Fig. 2b: if the large ship that is overtaking the sailboat, slowing down the sailboat will increase the risk of collision; in the (unlikely) situation of being the sailing boat overtaking the ship, reducing speed will only decrease *MDCPA* temporarily. In these cases, the strategy should be changing to the new course referred above until *MDCPA* measured with respect to the original course increases above the minimum. These two cases are considered when the angle difference between the absolute directions is below a certain threshold.

When the vessels are in crossing routes (Fig. 2c), slowing down the sailboat will only increase *MDCPA* if maintaining the current speed the sailboat already guarantees a passage behind the large ship. If keeping the current route the sailboat will pass in front if the large ship, then the strategy is to sail to the escape course until the large ship has passed and *MDCPA* increases above the safe threshold.

Although slowing down can be an effective way to reduce the risk of collision in the cases illustrated above, the ability to do that will depend strongly on the type of rig and the mechanism to regulate it, and the apparent wind angle. For example, a sailboat rigging a traditional two soft sail configuration with a shrouded mast will not be able to slow down for large apparent wind angles, when the boom is already pushing into the shrouds.

4.4 When to Initiate a Collision Avoidance Maneuver?

Not too late but also not too early. A late decision will naturally compromise the efficacy of the collision avoidance maneuver and deciding too early may introduce unnecessary disturbances in the mission being accomplished by the sailboat. This time will depend on the *TCPA*, *MDCPA* and also the estimated speed of the sailboat along possible escape courses to increase the minimum distance between the vessels. The practical rule is to consider only the approaching ships when *MDCPA* is smaller than a minimum safe radius *MSF* and *TCPA* is less than the time required to sail the distance *MSF* along the escape route COG_e: $TCPA < MSF/v_{ca}$ (where v_{ca} is the minimum between the current VMG and estimated VMG of the sailboat along the escape route COG_e). This guarantees that the maneuver to avoid the collision will start before crossing the route of the approaching ship.

Figure 3 illustrates two different scenarios that require sailing to a different route for avoiding a collision, showing the position where the decision to change route should occur. The sailboat navigates downwind (left) and upwind (right), the minimum safe distance *MSF* is equal to 0.5 nm, the downwind speed of the sailboat is 5 ktn and the VMG upwind equal to 2 ktn. In both cases the sailboat should initiate the collision avoidance maneuver when *TCPA* equals 15 min $= 0.5/min(2, 5) =$ 0.25 h, which guarantees enough time to navigate more than *MSF* away from the ship route.

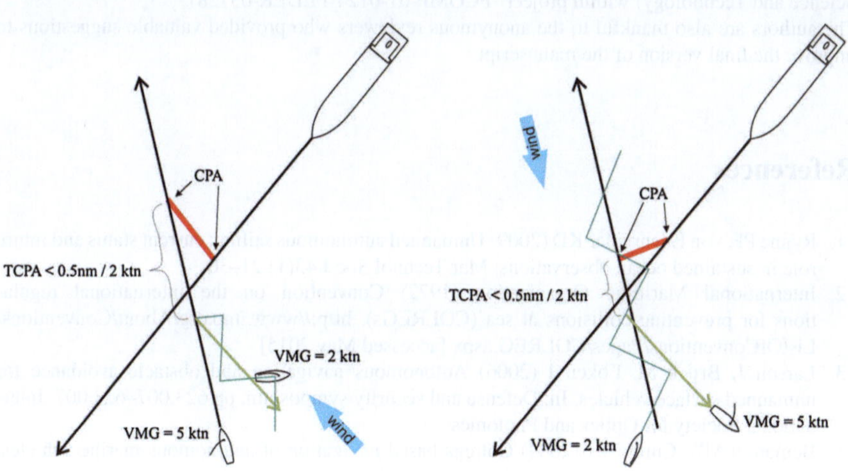

Fig. 3 Examples of points of decision for setting an escape route

5 Conclusions

Although the navigation rules established in the international regulations for preventing collisions at sea should guarantee a safe navigation for all ships adhering to them, there are many exception situations in the case of mismatch in maneuverability, size or observability of the vessels in a collision path. In long oceanic missions one real threat for small robotic sailing boats is the risk of collision with large ships that may not be able to change its route or speed for giving way, if the sailboat is identified too late or even ignored due to its small size. The basic set of rules proposed in this paper intends to establish a defensive and conservative behavior to avoid close encounters with AIS-transmitting vessels, promoting early maneuvers to reduce the risk of collision. Naturally this strategy does not make a robotic sailboat compliant with the international collision-avoidance rules, neither prevent other important hazards like collisions with small boats or entangling with floating debris for which other perception technologies are necessary. Even though this paper only considers a single ship at a time in route of collision, we will later extend the strategies presented here to take into account the identification of various ships in potential route of collision.

Acknowledgments This work is financed by the ERDF—European Regional Development Fund through the COMPETE Programme (operational programme for competitiveness) and by National Funds through the FCT—Fundação para a Ciência e a Tecnologia (Portuguese Foundation for Science and Technology) within project "FCOMP-01-0124-FEDER-037281".
The authors are also thankful to the anonymous reviewers who provided valuable suggestions to improve the final version of the manuscript.

References

1. Rynne PF, von Ellenrieder KD (2009) Unmanned autonomous sailing: current status and future role in sustained ocean observations. Mar Technol Soc J 43(1):21–30
2. International Maritime Organization (1972) Convention on the international regulations for preventing collisions at sea (COLREGs). http://www.imo.org/About/Conventions/ListOfConventions/Pages/COLREG.aspx [accessed May 2015]
3. Larson J, Bruch M, Ebken J (2006) Autonomous navigation and obstacle avoidance for unmanned surface vehicles. In: Defense and security symposium, pp 623,007–623,007. International Society for Optics and Photonics
4. Benjamin MR, Curcio JA (2004) Colregs-based navigation of autonomous marine vehicles. In: Proceedings of autonomous underwater vehicles
5. Naeem W, Irwin GW (2008) Evasive decision making in uninhabited maritime vehicles. In: Proceedings IFAC world congress, Milan, Italy, August, pp 12,833–12,838
6. Perera L, Carvalho J, Soares CG (2009) Autonomous guidance and navigation based on the colregs rules and regulations of collision avoidance. In: Proceedings of the international workshop advanced ship design for pollution prevention, pp 205–216
7. Statheros T, Howells G, Maier KM (2008) Autonomous ship collision avoidance navigation concepts, technologies and techniques. J Navig 61(01):129–142
8. Naeem W, Irwin GW, Yang A (2012) Colregs-based collision avoidance strategies for unmanned surface vehicles. Mechatronics 22(6):669–678

9. Kuwata Y, Wolf MT, Zarzhitsky D, Huntsberger TL (2014) Safe maritime autonomous navigation with colregs, using velocity obstacles. IEEE J Ocean Eng 39(1):110–119
10. United States Coast Guard, U.S. Department of Homeland Security: Automatic identification system overview. http://www.navcen.uscg.gov/?pageName=AISmain [accessed May 2015]

Consideration of State Representation for Semi-autonomous Reinforcement Learning of Sailing Within a Navigable Area

Hideaki Manabe and Kanta Tachibana

Abstract To sail quickly to a goal within a navigable area, complex control of the rudder and sail is required. Sailors must determine the current action with consideration of the time series of states; i.e., both current and future states. Reinforcement learning is an appropriate method for learning a complex problem, such as sailing. In this paper, we apply the navigable area such that a robotic sailor must avoid touching a boundary. To realise a higher layer of sailing architecture, the action space is simplified and discretised to the degree of the sailboat direction change. Moreover, we utilize semi-autonomous reinforcement learning, also known as imitation learning, in which a human selects an action and a robot updates its Q-values to evaluate pairs of states and actions until the robot's action selection is equivalent to the human's. For semi-autonomous learning, as well as for normal reinforcement learning, a representation of the state space is important. The state representation should be defined so that the state space is discretised to specify a desirable action, thereby removing any redundancy if possible. In this paper, we verify and investigate the possibility of state representation.

1 Introduction

Reinforcement learning is a machine learning method proposed by Sutton and Barto [6]. Agents of learning recognize states and select actions. They learn how each state and action pair contributes to the rewards. The objectives of the World Robotic Sailing Championship consist of sub-tasks to move the sailboat to the desired area.

H. Manabe (✉)
Laboratory for Intelligence, Kogakuin University, Nishi-Shinjuku, Shinjuku, Tokyo, Japan
e-mail: em14012@ns.kogakuin.ac.jp

K. Tachibana (✉)
Laboratory for Intelligence, Faculty of Informatics, Kogakuin University, Nishi-Shinjuku, Shinjuku, Tokyo, Japan
e-mail: kanta@cc.kogakuin.ac.jp

A. Friebe and F. Haug (eds.), *Robotic Sailing 2015*,
DOI 10.1007/978-3-319-23335-2_7

The robot must recognize continuous state variables, position (x, y) and velocity $(\dot{x}, \dot{y})$, and determine the appropriate action to the destination. However, unlike the control of driverless cars, the acceleration of a sailboat depends on the apparent wind. Therefore, we make the agents learn to navigate to the goal using reinforcement learning. In addition, we consider that state is better represented by the directions of the apparent wind, goal areas, and obstacles in the boat coordinate system than by the sailboat position and velocity in the global coordinate system. The master's dissertation of Sterne [5] realised this observation by sailing to a desired direction through reinforcement learning. However, neither the task of avoiding obstacles nor route finding in navigable areas was realised.

Konidaris and Barto [1] produced successful results of reinforcement learning experiments in the pinball domain in which agents learned the route to a goal using bounces to fixed elastic walls. In the experiment, the state vector of continuous values $(x, y, \dot{x}, \dot{y})$ was used. They employed agents to choose an action out of five possible actions ($\dot{x} \leftarrow \dot{x} \pm \Delta, \dot{y} \leftarrow \dot{y} \pm \Delta$ or no accelerations, where Δ is a small fixed value). In our sailing task, we make sailboats circumnavigate a target area without touching the navigable area boundary by using continuous state spaces. State spaces should be represented in the sailboat coordinate system because the danger of colliding with a boundary of the navigable area depends on the apparent wind, even if the relative position and velocity moving toward the boundary of the navigable area are the same.

The number of state variables becomes too large if all measured variables are used when the agent is made to recognize the direction and distance to obstacles and the target area. The number of state variables can be reduced if we use the Fourier transform of measured distance data. Kuhl [2] showed that the original distance signal can be restored by inverse Fourier transformation with a relatively small number of Fourier coefficients.

In this paper, we simplify the action spaces, utilize semi-autonomous learning, and arrange state representation to make the agent circumnavigate target areas within the navigable area. Action space is discretized to $\Delta\theta \in \{0°, \pm 1°\}$, not to continuous values, as was done in Sterne's dissertation. We adopt semi-autonomous reinforcement learning which is also known as imitation learning [3, 4]. Semi-autonomous reinforcement learning is the method of learning actions that are selected by the human's operation at the beginning stage of learning, and the robot learns from the human's actions. For state representation, distances to the target area and boundaries to the navigable area at respective directions are measured in the sailboat coordinate system. They are then Fourier-transformed to reduce the number of state variables. The state variables consist of Fourier coefficients of distance to the target area and to obstacles, and a two-dimensional vector of apparent wind. We define the relative direction θ with respect to the sailboat. Then, we calculate distances to the target area $[d_+(\theta)]$ for all angles $\theta = 0°, 1°, \ldots, 359°$ around the sailboat. We calculate the distances to the boundaries of the navigable area $[d_-(\theta)]$ in the same way. If a ray to angle θ does not cross the target area, $d_+(\theta)$ becomes ∞. We define $\ell = \exp(-d)$ so that ℓ is within the range from 0 to 1. The functions $\ell_+(\theta)$ and $\ell_-(\theta)$ are naturally periodic; therefore, they are

Fourier-transformed. In this paper, we verify whether a reduced number of Fourier coefficients effectively represent the state space combined with the apparent wind vector.

The remainder of this paper is organized into six sections. In Sect. 2, sailing and our sailing simulator are explained. In Sect. 3, semi-autonomous learning and Q-learning are described. In Sect. 4, state space is defined and the methods of performing Fourier transformation of state representation and the dividing of state space are respectively described. In Sect. 5, our experiment and parameters, experimental scenario, and results are presented. In Sect. 6, we discuss our experiment. Our research conclusions and future work are discussed in Sect. 7.

2 Sailing Simulator

The main propulsive force of the sailboat is captured by the sail. This force can be reasonably approximated as being proportional to the sail area facing the wind and the squared wind speed. The propulsive component of this force pushes the sailboat forward, whereas the lateral component moves the sailboat sideways. Let W be the apparent wind speed, ϕ be the direction of the apparent wind, and φ' be the direction of the sail. Then, the sail gains force such that

$$F \propto W^2 \sin(\phi - \varphi'),$$

and its propulsive and lateral components are

$$F_x \propto W^2 \sin(\phi - \varphi') \sin \varphi',$$
$$F_y \propto W^2 \sin(\phi - \varphi') \cos \varphi',$$

respectively. For a given W and ϕ, the half angle $\varphi' = \phi/2$ maximizes the propulsive force, F_x. Therefore, $F_x \propto W^2(1 - \cos\phi)$ and $F_y \propto W^2 \sin\phi$

Figure 1 presents a screenshot of the sailing simulator used in this study. The wind direction, φ, is set to East-North-East; i.e., at $\varphi = -22.5°$ in the world coordinate system. The 'dead zone' is indicated by a striped pattern. When steering in the dead zone, the direction of the apparent wind becomes almost $\phi \approx 0°$; therefore, the propulsive force component is lost, and it is difficult for the sailboat to gain propulsion. Accordingly, sailors heading upwind cannot move directly and must choose a zigzag course to arrive at their goal in a shorter time. The direction perpendicular to the wind is called the 'abeam', in which the sailboat can move fast by maintaining a high apparent wind speed, W, and a good propulsive component $(1 - \cos\theta)$. Finally, the downwind direction is called the 'running', in which the

Fig. 1 Screenshot of the sailing simulator

sailboat cannot go faster than the abeam because the true wind is cancelled by the speed of the sailboat, and the apparent wind becomes approximately $W \approx 0$. Therefore, to arrive at the goal in a shorter time, the sailboat must avoid the dead zone and running directions and maintain an abeam direction as much as possible.

In the sailing simulator, it is assumed that the resistance force from the water is proportional to the square of the boat velocity. The leeway effect is also simulated with consideration of the lateral motion equation. Moreover, if the sailboat touches a boundary of the navigable area, its velocity is immediately changed to zero. The frame rate of the sailing simulator is set to 40 frames per second.

3 Semi-autonomous Reinforcement Learning

In Sect. 3.1, we explain reinforcement learning, especially Q-learning. We explain the method to renew Q-values and to select the action. In Sect. 3.2, we propose the method of semi-autonomous reinforcement learning.

3.1 Reinforcement Learning

Q-learning is a reinforcement learning method. An agent has a finite set of states S and a finite set of actions A. At each time frame, t, an agent recognizes its state, $s_t \in S$, and selection action, $a_t \in A$. The efficacy of a pair of states and actions is quantified as the 'Q-value', $\hat{Q}(s_t, a_t)$. The Q-value is renewed with obtained reward r_t and Q-values of the next state s_{t+1} caused by action a_t as the following:

$$\hat{Q}(s_t, a_t) \leftarrow (1-\alpha)\hat{Q}(s_t, a_t) + \alpha\left\{r_t + \gamma \max_a \hat{Q}(s_{t+1}, a)\right\},$$

where $\alpha \in [0, 1]$ and $\gamma \in [0, 1]$ are the learning rate and discount rate, respectively. The method to select the action with the highest Q-value for the current state is called the greedy method. The method to select the action according to the greedy method with the probability of $(1-\varepsilon) \in (0, 1)$, and to otherwise randomly select the action, is called the ε-greedy method. The soft-max method is used to select action a with a probability that is proportional to $\exp(\beta\hat{Q}(s_t, a))$, the exponential of its Q-value, where β is a constant. Q-value is updated for each step. Q-values are set at 0 initially.

3.2 *Semi-autonomous Reinforcement Learning*

We propose a semi-autonomous reinforcement learning technique. In semi-autonomous learning, a human operator teaches robots the best route to the target area. Robots share human decision-making through Q-values, which are renewed with each frame in the same way as in conventional Q-learning.

4 State Representation

To find an appropriate route to a target area within a navigable area, the robot must recognize its state and properly select the action. Recognition of the state is important; it depends on representation of the state space. We investigate the following state representation consisting of Fourier coefficients of distances to the target area, Fourier coefficients of distances to the boundaries of the navigable area, and the two-dimensional vector of apparent wind.

We define the relative direction, θ, with respect to the sailboat. We assume that agents can precisely measure their Euclidean distances to the target area $[d_+(\theta)|\theta = 0°, 1°, \ldots, 359°]$ and the boundaries of the navigable area $[d_-(\theta)|\theta = 0°, 1°, \ldots, 359°]$ for each degree. If a ray to angle θ does not cross the target area, $d_+(\theta)$ is set to ∞. The distance functions are transformed to 'nearness' functions, $\ell_+(\theta) = [e^{-d_+(\theta)}]$ and $\ell_-(\theta) = [e^{-d_-(\theta)}]$, respectively. Then, the periodic nearness functions are Fourier-transformed, respectively. The number of original nearness signals is 360 for each function. After Fourier transformation:

$$\ell(\theta) = a_0 + a_1 \cos\theta + b_1 \sin\theta + a_2 \cos 2\theta + b_2 \sin 2\theta + \cdots + a_N \cos N\theta$$
$$+ b_N \sin N\theta + \ldots,$$

$$a_0 = \frac{1}{360} \sum_{\theta=0^\circ}^{359^\circ} f(\theta)$$

$$a_k = \frac{1}{720} \sum_{\theta=0^\circ}^{359^\circ} \ell(\theta) \cos k\theta, b_k = \frac{1}{720} \sum_{\theta=0^\circ}^{359^\circ} \ell(\theta) \sin k\theta, k \in \{1, \ldots, N\},$$

we use $(2N+1)$ coefficients of N lowest wave numbers, i.e., $a_0, a_1, b_1, \ldots, a_N, b_N$, as state variables. In addition, we assume that the sailboat precisely measures the apparent wind. The x, y components of the apparent wind in the sailboat coordinate system are used as other state variables. We investigate the state space of $(2N_+ + 1) + (2N_- + 1) + 2$ dimensions in total, where N_+ and N_- are wave numbers for $\ell_+(\theta)$ and $\ell_-(\theta)$, respectively. We scale apparent wind and a_0 of both distances by 0.5, a_k and $b_k, k \in \{1, \ldots, N_+\}$ and $k \in \{1, \ldots, N_-\}$ by 2 to adjust Euclidean distance in the state space.

Figure 2 shows an example of sailboat state. The target area is from 45 degrees left to 45 degrees right in the front direction of the sailboat. The shortest distance is $d_+(0°) = 300$. Also, the shortest distance to the boundary of navigable area is $d_-(270°) = 300$. Figure 3 shows $d_+(\theta)$, $d_-(\theta)$, $\ell_+(\theta)$ and $\ell_-(\theta)$ respectively.

The continuous state space is partitioned to m regions. Before semi-autonomous learning, a human operator makes the sailboat travel in the navigable area and stores state data for each frame. Then, m state data are randomly picked up and used as generators of Voronoi division. While learning, the robot must discretize its continuous state vector. For each frame, the robot calculates a state vector and recognizes it as one of the m states corresponding to the nearest generator in the Euclidean distance.

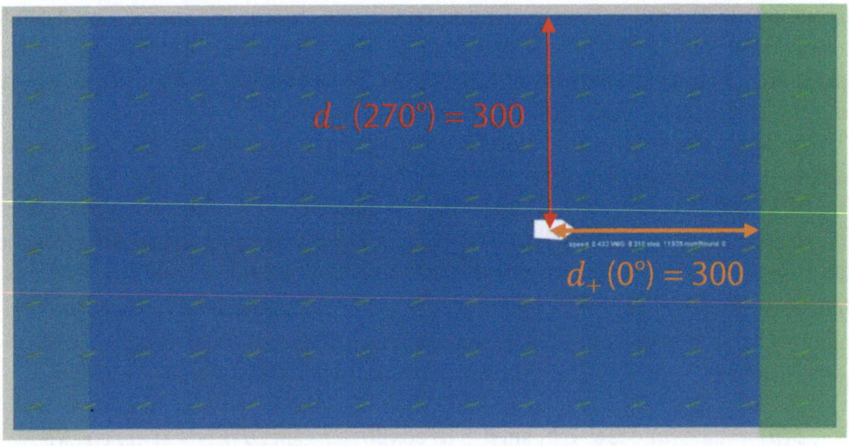

Fig. 2 An example of sailboat state

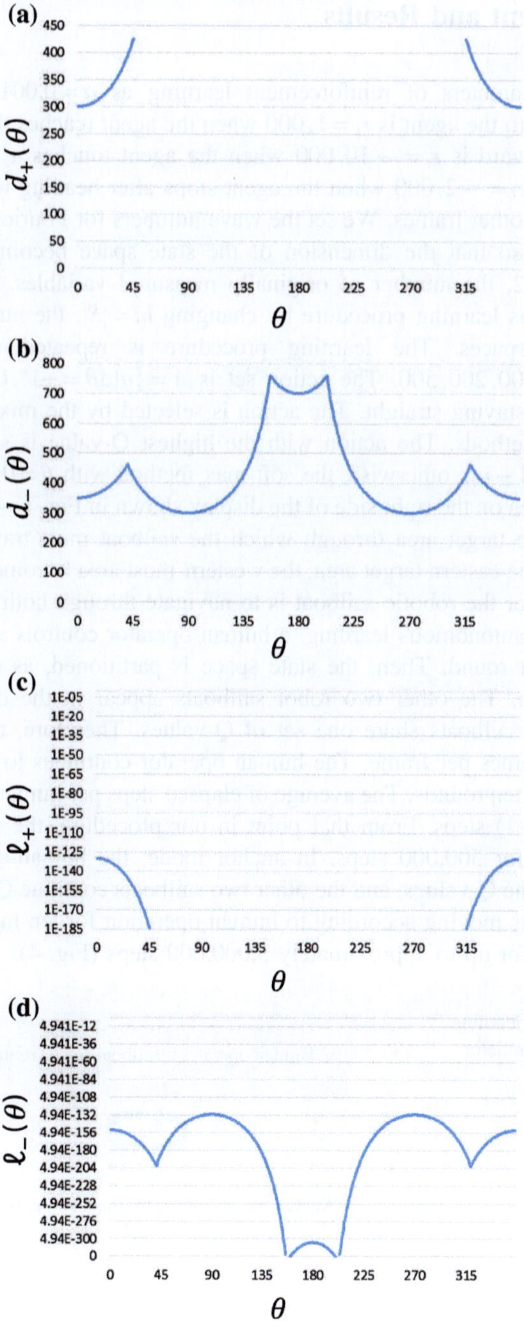

Fig. 3 The distances to boundaries of navigable area $d_+(\theta)$ and $d_-(\theta)$, the nearness $\ell_+(\theta)$ and $\ell_-(\theta)$ **a** The distance to target area $d_+(\theta)$ **b** The distance to boundaries of navigable area $d_-(\theta)$ **c** The nearness $\ell_+(\theta)$ **d** The nearness $\ell_-(\theta)$

5 Experiment and Results

We set the parameters of reinforcement learning as $\alpha = 0.001$ and $\gamma = 0.999$. A reward given to the agent is $r_t = 1,000$ when the agent reaches the target area. In addition, the reward is $r_t = -10,000$ when the agent touches a boundary of the navigable area; $r_t = -2,000$ when the agent stops after heading to the dead zone; and $r_t = -1$ for other frames. We set the wave numbers for Fourier transformation, $N_+ = N_- = 17$, so that the dimension of the state space becomes 72, which is one-tenth of 722, the number of originally measured variables. We execute the semi-autonomous learning procedure by changing $m = |S|$, the number of discretized state subspaces. The learning procedure is repeated twice for each $m = 10, 20, 50, 100, 200, 500$. The action set is $A = \{a | \Delta\theta = \pm 1°, 0°\}$; i.e., turning right or left, or staying straight. The action is selected by the mixture of ε-greedy and soft-max methods. The action with the highest Q-value is selected with the probability of $(1 - \varepsilon)$; otherwise, the soft-max method with $\beta = 0.1$ is applied.

The green area on the right side of the display shown in Fig. 1—for example, the east side—is the target area through which the sailboat must traverse. When the sailboat enters the eastern target area, the western-most area becomes the next target area. The task for the robotic sailboat is to navigate through both target areas.

Before semi-autonomous learning, a human operator controls a sailboat until it finishes the first round. Then, the state space is partitioned, as described in the previous section. The other two robot sailboats appear in the display and start learning. Three sailboats share one set of Q-values. Therefore, the Q-values are renewed three times per frame. The human operator continues to control the sailboat for the first ten rounds. The average of elapsed steps per human-operated round is 1,424.8 (±36.1) steps. From that point in our procedure, the sailboat turns to 'anchor mode' for 500,000 steps. In anchor mode, the human-operated sailboat stops updating the Q-values, and the other two sailboats continue Q-learning. Then, the sailboat starts moving according to human operation for ten more rounds. This loop continues for up to approximately 3,000,000 steps (Fig. 4).

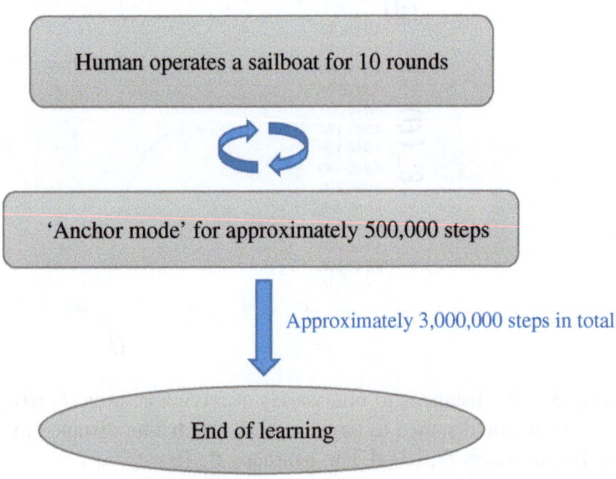

Fig. 4 Process of learning with the sailing simulator

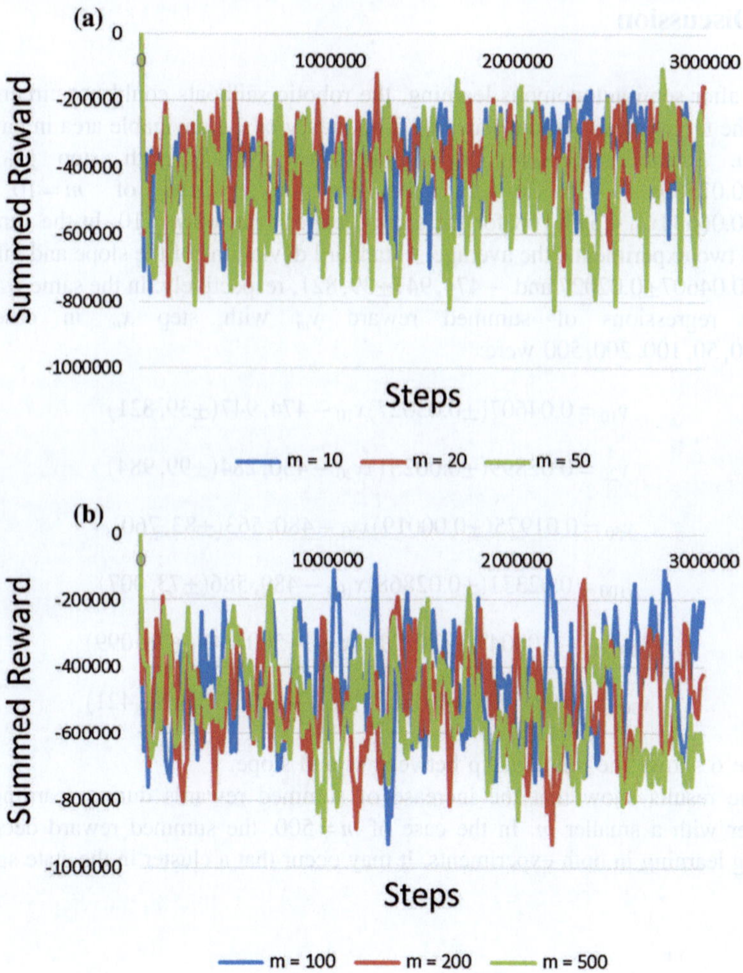

Fig. 5 Steps and summed reward **a** Reward in the case in which $m = 10, 20, 50$ **b** Reward in the case in which $m = 100, 200, 500$

Figure 5 shows the summed rewards in the cases where $m = 10, 20, 50$, and $m = 100, 200, 500$. 'Summed reward' means the summation that rewards the three sailboats, which are one's own sailboat and the agents' two sailboats 10,000 step. The X axis is the number of steps in the learning; the Y axis is the summed reward in the learning.

6 Discussion

Even after semi-autonomous learning, the robotic sailboats could not circumnavigate the target areas without touching a boundary of the navigable area in any case of m. Linear regression of summed reward y_{10} with step x_{10} was $y_{10} = 0.02580x_{10} - 435,126$ for the first experiment of $m = 10$, and $y_{10} = 0.06634x_{10} - 514,768$ for the second experiment of $m = 10$. In the summary of the two experiments, the average ± standard deviations of the slope and intersect were 0.04607±0.02027 and −474,947±39,821, respectively. In the same way, the linear regressions of summed reward y_m with step x_m in cases of $m = 20, 50, 100, 200, 500$ were:

$$y_{10} = 0.04607(\pm 0.02027)x_{10} - 474,947(\pm 39,821)$$

$$y_{20} = 0.02899(\pm 0.00251)x_{20} - 430,284(\pm 99,984)$$

$$y_{50} = 0.01975(\pm 0.00019)x_{50} - 480,563(\pm 83,760),$$

$$y_{100} = 0.02331(\pm 0.02868)x_{100} - 489,586(\pm 73,007)$$

$$y_{200} = -0.00047(\pm 0.03219)x_{200} - 500,849(\pm 78,099)$$

$$y_{500} = -0.06096(\pm 0.02676)x_{500} - 440,724(\pm 80,421)$$

Figure 6 shows the relationship between m and slope.

The results show that the increase of summed rewards during learning was greater with a smaller m. In the case of $m = 500$, the summed reward decreased during learning in both experiments. It may occur that a cluster in the state space is

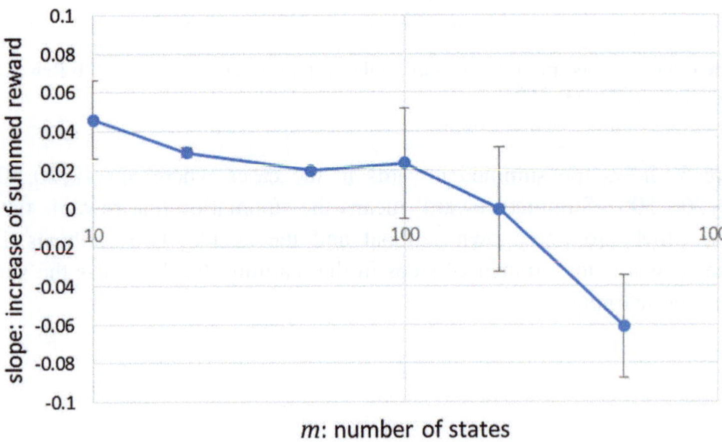

Fig. 6 Graph of approximations to steps and all rewards in Fig. 5

subdivided to too many partitions in the cases of larger m. If similar states are discretized to different state subspace and recognized as different discrete state, more updates of Q-values are needed than appropriate number of partitions. We compare progress of learning by action selection entropy between the cases of $m = 10$ and $m = 500$.

We calculate action selection entropy for each state. If action selection entropy is small for a state, a robot always selects the same action in the state. The minimum of action selection entropy is 0, where the probability of an action is selected is 1 and other actions are not selected. And if action selection entropy is large, a robot selects actions at random. The maximum is $\log_2 3$ in our case, where each of three possible actions is selected with the probability of one third. Action selection entropy for a state s is:

$$H(s) = -\sum_{a \in A} P(a|s),$$

$$P(a|s) = \frac{\exp\left(\beta \hat{Q}(s,a)\right)}{\sum_{a' \in A} \exp\left(\beta \hat{Q}(s,a')\right)},$$

We denote the set of states chosen for the experiment of $m = 10$ as S_{10}. We calculate action selection entropy for each state in S_{10} to find the state s^* with the minimal entropy after the learning procedure. Then, we find subset of states:

$$S^*_{500} = \left\{ s \in S_{500} | s^* = \operatorname*{argmin}_{s' \in S_{10}} d(s,s') \right\},$$

here S_{500} is the set of states chosen in the experiment of $m = 500$, and $d(s,s')$ is the Euclidean distance between states s and s'. The subset S^*_{500} consists of states that are in the Voronoi region of s^*. Action selection entropy is calculated for each state of S^*_{500}.

Four combinations are evaluated, i.e. s^* is found for each experiment of $m = 10$, and S^*_{500} is detected for each experiment of $m = 500$. $H(s^*)$ was 1.200 for the first experiment of $m = 10$. $|S^*_{500}|$ was 64 and 76 for the first and the second experiments, respectively. $H(s^*)$ was 1.268 for the second experiment of $m = 10$. $|S^*_{500}|$ was 76 and 76 for the first and the second experiments, respectively. Figure 7 shows histograms of entropy in S^*_{500} for each experiment. $H(s^*)$ are shown by red arrows for each experiment. Majority of S^*_{500} have larger entropy than s^* and action is selected almost at random in such states.

However, the robotic sailboats learned the action of avoiding the obstacles. For example, we showed state 39 as one of 100 states. The red dot in Fig. 8 denotes where the sailboat recognized its state as state 39. The agent selected the action to turn left in state 39. Figure 9 shows the distance to the goal area and to the boundaries of the navigable area restored by inverse Fourier transformation from the state vector of state 39. The solid and dotted lines show restored $\ell_+(\theta)$ and $\ell_-(\theta)$, respectively. The direction of apparent wind is shown by the arrow.

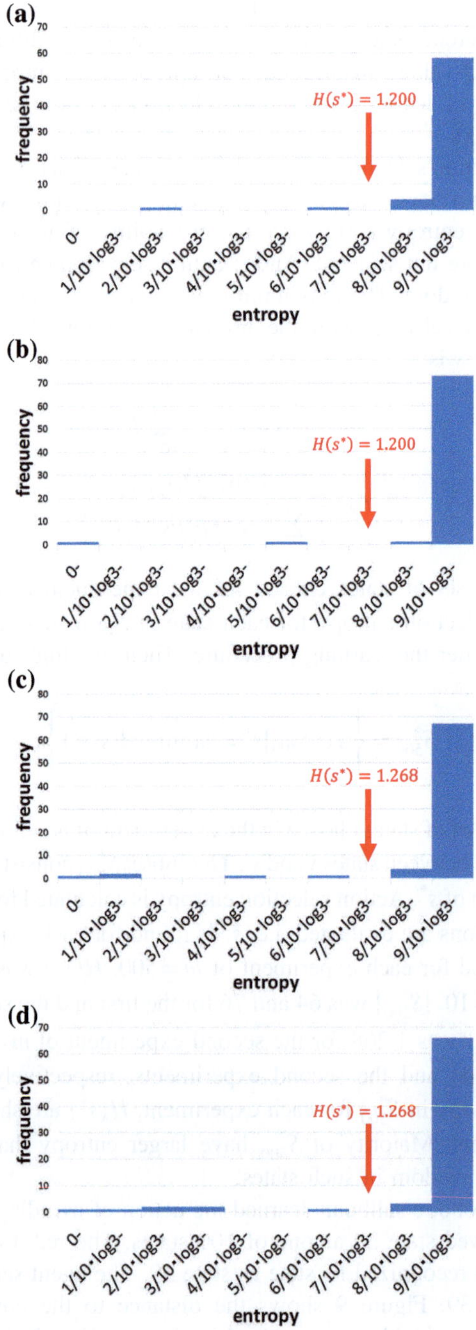

Fig. 7 Histograms of entropy **a** Histogram of entropy in S^*_{500} for the first experiment and $H(s^*)$ for the first experiment **b** Histogram of entropy in S^*_{500} for the second experiment and $H(s^*)$ for the first experiment **c** Histogram of entropy in S^*_{500} for the first experiment and $H(s^*)$ for the second experiment **d** Histogram of entropy in S^*_{500} for the second experiment and $H(s^*)$ for the second experiment

Fig. 8 Screenshot in state 39 at the number of 100 state spaces

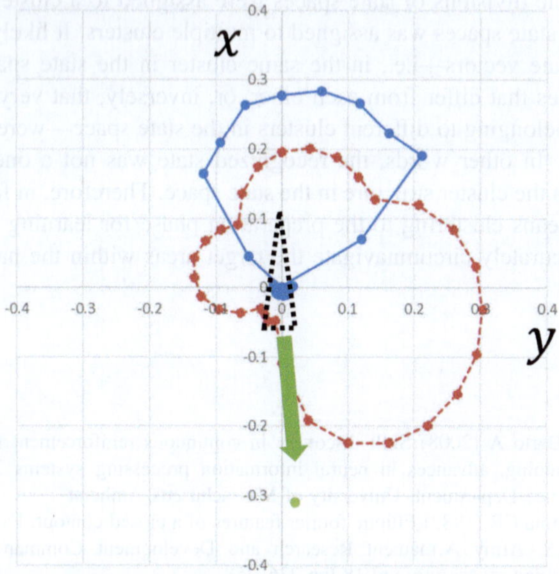

Fig. 9 Value of $\ell_+(\theta)$ and $\ell_-(\theta)$ and wind direction at state 39

The target area was in the forward direction of the sailboats. The nearest boundary of the navigable area was on the right side of the sailboats. The direction of apparent wind was forward and somewhat left in state 39. In addition, the goal area was located ahead or somewhat to the left, and the boundaries of the navigable area were located on the right of the sailboats. The direction of wind was forward in state 39 of Fig. 6; the state was certainly recognized in the results. The Q-values at state 39 were -711.9 (turn left), -717.9 (go straight), and -720.0 (turn right); therefore, the sailboats certainly learned to avoid the boundary by turning left. From this result, we can see that the Fourier coefficients of $N_- = 17$ of the lowest wave numbers stored sufficient information to avoid touching the boundaries.

7 Conclusion

In this paper, we described previous research of reinforcement learning for control in continuous state space in introduction. We explained our sailing simulator, and proposed a semi-autonomous reinforcement learning technique. We did the experiments at each number of divisions of the state space. As the result, the sailboat did not navigate to the goal area in these experiments. Nevertheless, the agent employed the action that avoided the boundaries of the navigable area when they were approached. We calculated the value of $\ell_+(\theta)$ and $\ell_-(\theta)$ to use inverse Fourier transformation from Fourier coefficients. We confirmed that the states were certainly recognized.

The state representations were calculated as the paths of the sailboats first approaching the navigable area, and they were chosen at random. In this division method, multiple divisions of state spaces were assigned to a cluster or, inversely, the division of state spaces was assigned to multiple clusters. It likely occurred that very similar state vectors—i.e., in the same cluster in the state space—were recognized as states that differ from each other, or, inversely, that very different state vectors—i.e., belonging to different clusters in the state space—were recognized as the same state. In other words, the recognized state was not a one-to-one correspondence with the cluster structure in the state space. Therefore, in future work, we will apply *k*-means clustering to the preparation phase for learning for the robotic sailboats to accurately circumnavigate the target areas within the navigable area.

References

1. Konidaris G, Barto A (2008) Skill discovery in continuous reinforcement learning domains using skill chaining, advances in neural information processing systems 22 (NIPS 2009), Computer Science Department, University of Massachusetts Amherst
2. Kuhl FP, Giardina CR (1982) Elliptic fourier features of a closed contour, Fairleigh Dickinson University, U.S. Army Armament Research and Development Command. In: Computer graphics and image processing, vol 18, pp 236–258
3. Ross S, Gordon GJ, Bagnell JA (2011) A reduction of imitation learning and structured prediction to no-regret online learning. In: Appearing in proceedings of the 14th international conference on artificial intelligence and statistics (AISTATS), JMLR: W&CP 15, vol 15. Fort Lauderdale, FL, USA
4. Shon AP, Verma D, Rao RPN (2007) Active imitation learning, AAAI conference on artificial intelligence, Department of Computer Science and Engineering, pp 756–762
5. Sterne PJ (2004) Reinforcement sailing [master's thesis], Master of science. School of Informatics, University of Edinburgh
6. Sutton RS, Barto AG (1998) Reinforcement learning, a bradford book. The MIT Press, Cambridge

AEOLUS, the ETH Autonomous Model Sailboat

Jonas Wirz, Marco Tranzatto, Alex Liniger, Marcello Colombino, Henrik Hesse and Sergio Grammatico

Abstract Path planning and control are particularly challenging tasks for a sailboat. In contrast to land vehicles or motorboats, the movement of a sailboat is heavily restricted by the wind direction. This paper focuses on the low-level control acting on the rudder and the sails. Specifically, a standard proportional controller and a non-linear controller have been implemented to track a reference heading. Further, special control algorithms that are activated during a tack or a jibe perform fast and smooth maneuvers. The path planner is based on the minimization of the weighted sum of different cost functions and allows for multi-objective optimization of the boat trajectory such as obstacle avoidance, time-to-target minimization and tactical behaviors.

1 Introduction

This paper reports about the development of an autonomous model sailboat of the Automatic Control Laboratoy at ETH Zurich, called *Aeolus*, shown in Fig. 1. Specifically, we address three main tasks to enable autonomous operation of model sailboats: sailing with a fixed heading, executing a fast and smooth tack maneuver and

J. Wirz · M. Tranzatto · A. Liniger · M. Colombino · H. Hesse(✉) · S. Grammatico(✉)
Automatic Control Laboratory, ETH Zurich, Switzerland
e-mail: grammatico@control.ee.ethz.ch

H. Hesse
e-mail: hesseh@control.ee.ethz.ch

M. Tranzatto
e-mail: marco.tranzatto@gmail.com

A. Liniger
e-mail: liniger@control.ee.ethz.ch

M. Colombino
e-mail: mcolombi@control.ee.ethz.ch

J. Wirz
e-mail: wirzjo@student.ethz.ch

A. Friebe and F. Haug (eds.), *Robotic Sailing 2015*,
DOI 10.1007/978-3-319-23335-2_8

Fig. 1 Aeolus sailing on the Lake Zurich, Switzerland. The boat is an international one-meter RC model sailboat with specialized electronics installed. At the bow we placed the weather station AIRMAR WS-200WX [1]. The boat was used for all field tests and is equipped with a Pixhawk autopilot [2] providing data logging and telemetry

planning a path in presence of obstacles. The path planning controller proposes reference actions to the low-level control, namely: *track a reference heading*, *tack* and *jibe*.

We started with an international one-meter RC model sailboat shown in Fig. 1, and installed specialized electronics. This setup is described in [3]. The main information used to sail and navigate are provided by an extended Kalman filter (EKF) already integrated into the firmware of the Pixhawk autopilot [2], which is our main micro controller unit. The EKF supplies the heading angle, ψ, while the weather station measures the apparent and estimated true wind direction, σ, speed, v, as well as the GPS position (latitude, longitude and altitude) and the course over ground (COG), χ, as defined in Fig. 2a, b. Due to drift of the boat the heading angle and the COG are typically not identical.

We design two moving average filters (window length 2 s) for the raw measurement of the direction and speed of the true wind to filter out high-frequency wind shifts [3]. This would negatively influence the heading angle with respect to the wind direction, which reads as $\alpha = \psi - \sigma$.

Fig. 2 **a** Heading angle ψ and the course-over-ground angle χ. v is the velocity vector of the vessel. **b** the relative heading angle α: ψ is the heading of the boat and σ defines the direction of the true wind

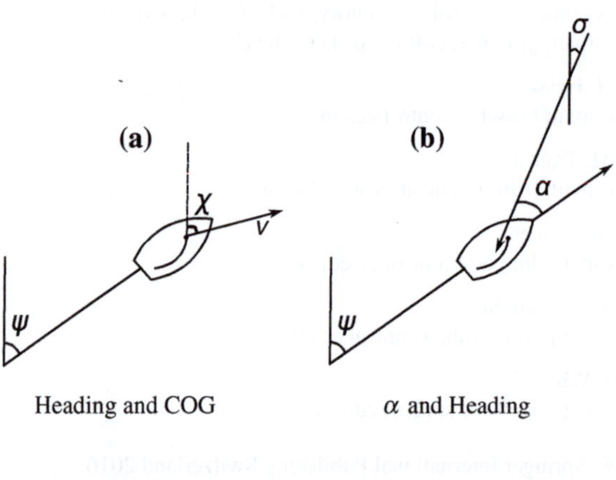

To overcome drift due to currents and waves, we use the course over ground value χ, provided by the GPS signal, instead of the heading ψ. Because the GPS signal sometimes drops, even for long periods of time, up to 10 s, we propose to use an estimated $\hat{\alpha}$ angle, which is a convex combination between $\alpha_\psi = \psi - \sigma$ and $\alpha_\chi = \chi - \sigma$, as follows:

$$\hat{\alpha} = (1-\lambda)\alpha_\chi + \lambda\alpha_\psi, \quad \lambda = \begin{cases} 1 & \text{if } t - t_{\text{last}} > t_{\text{threshold}} \\ \frac{t-t_{\text{last}}}{t_{\text{threshold}}} & \text{otherwise} \end{cases}, \tag{1}$$

where t is the current time, t_{last} is the time when the last update of the COG was available, and $t_{\text{threshold}}$ is a selectable threshold time (1.5 s).

To tune and analyze the closed-loop behavior induced by the controllers, a dynamical model for the sailboat is required. The dynamic discrete-time linear system $x_{k+1} = Ax_k + B\delta_k$, where δ is the rudder command and $A \in \mathbb{R}^{2\times2}$ and $B \in \mathbb{R}^{2\times1}$, describes the evolution of the state vector $x_k = \left[\omega_k, \psi_k\right]^\top$, where ψ is the yaw and ω the yaw rate. We identified the state space model from several lake tests at the Lake Zurich, Switzerland, by using step commands in the rudder. This model has been validated using data collected in several navigation tests. We refer to [3] for a discussion on the identification results.

2 Low-Level Control

A low-level regulator is in charge of controlling both the sails and the rudder separately. In literature, fuzzy controllers [4–6], standard approaches with PI control [7, 8] or a nonlinear approach with back stepping [9] were used to regulate the rudder of a sailboat. The reference heading angle for the low-level regulator is $\alpha^\star$, and is set by the medium-level controller. Using the equation $\alpha = \psi - \sigma$, we can specify either a constant compass course or a constant heading to the wind as reference. The first case is simply obtained by setting $\sigma = 0$; the second case uses a more general formulation where σ is provided by sensor measurements.

Here we design two rudder controllers: a standard proportional (P) controller and a more sophisticated nonlinear one (NL). Let us normalize the rudder command δ to be in the range $[-1, 1]$. Given the reference angle $\alpha^\star$ and the current estimate α, the heading error reads as $e := \alpha^\star - \alpha$, and the P rudder controller sets the rudder command δ as $\delta = \delta_{\text{P}}(e) = k_{\text{p}}\, e, k_p > 0$.

Instead, the NL controller defines a nonlinear gain $k(e)$ as $k(e) = \frac{k_{\text{p}}}{1+c_{\text{p}}|e|}$ for some $k_{\text{p}}, c_{\text{p}} > 0$, and hence sets the rudder command to $\delta = \delta_{\text{NL}}(e) = k(e)\, e$.

The NL controller was first introduced in [10], it acts as a proportional gain when the error e is small, but its behavior changes when the error is large. The c_{p} constant can be in fact used to tune the control action when the error is large; here we tune the

NL controller such that a large c_p leads to a smaller gain $k(e)$ and resulting rudder action.

After several experimental tests, we have tuned k_p for both stability and tracking purposes to the value $k_p = 0.35$. Based on the results from the P controller, we can tune the NL one with $k_p = 0.35$, $c_p = 0.35$ [3].

As for the sail control, we combine ideas from rule-based control and extremum seeking sail control (ESSC) [11]. For low speeds (<0.5 m/s), a simple rule based linear law is used: the more Aeolus is sailing opposite to the wind direction, i.e. the smaller α, the more we close the sails. For speeds above the threshold the "trial and error" approach ESSC is used that tries to maximize the speed of the boat by altering the sail angle. The details of the sail controller are not discussed further in this paper.

During a tack the boat crosses the no-sail zone, therefore the maneuver should be executed in the fastest and smoothest possible way using the inertia of the boat to push it through the wind.

From our field experience [3], we suggest three critical actions to be taken into account when tacking. First, the window size of two average filters (on the wind direction σ and on the α angle) is set to one sample only; second, λ in Eq. (1) is set to 1; third, a specialized tack regulator takes control of the rudder during the maneuver. The first action overcomes the undesired delay caused by the filters. The second action is useful because during a fast tack maneuver, it is likely to loose the updated COG measurement. Three possible ways of carrying out a tack are developed and tested: the *implicit*, the *dedicated*, and the *LQR* one [3].

In this paper we focus on the LQR tack controller. It optimizes the tack maneuver with respect to an infinite-horizon cost function. The LQR is a state-feedback controller, meaning that the control action is a static feedback law of the state, x that is $\delta = \delta(x) = Kx$.

To exploit the model derived in [3], we make the assumption that the true wind direction does not change during the tack maneuver, which is a practically reasonable assumption if the maneuver is fast enough. In this way, a tack maneuver can be just seen as a change in the heading angle ψ. Thus, tacking results in steering the state of the system, x_k, from the initial value $x_i = [\omega_i, \psi_i]^T$ to the final value $x_f = 0$, where the latter implies achieving the desired $\alpha^\star$ with zero yaw rate.

Let us obtain the state space formulation of our system. We define the variables

$$\hat{\delta}_k := \delta_k - \delta_{k+1} \qquad \hat{x}_k := [\omega_k, \psi_k, \delta_{k-1}]^T, \tag{2}$$

where $\hat{\delta}$ is the new control input and $\hat{x} \in \mathbb{R}^3$ is the extended state. The main reason to extend the state vector as in Eq. (2) is to assign a cost penalty to the control action δ as well as to the control variation $\hat{\delta}$ [3]. Namely, the extended state at time k, $\hat{x}_k$, contains the yaw rate and yaw angle at time k and the rudder command δ_{k-1} injected into the system at the previous step $k-1$. The input $\hat{\delta}$ is the difference between the actual rudder command at time k, and the previous one; in other words, the real rudder command δ provided at the time k is then $\delta_k = \delta_{k-1} + \hat{\delta}_k$.

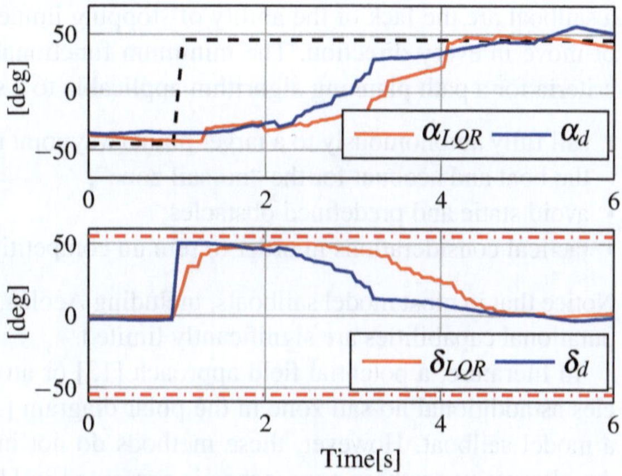

Fig. 3 Comparison between LQR and dedicated tack from port to starboard haul. The *upper plot* shows the reference $\alpha^\star$ (*dashed black*), the LQR (*solid orange*) and the dedicated (*solid blue*) tack responses. The *lower plot* shows the rudder limits (*dotted red*), the LQR (*solid orange*) and the dedicated (*solid blue*) rudder commands. The LQR tack takes ~3.2 s, while the dedicated one takes ~3.7 s

The state space matrices $\hat{A}$ and $\hat{B}$ corresponding to the extended state dynamics become

$$\hat{A} := \begin{bmatrix} A & B \\ 0 & I \end{bmatrix}, \; \hat{B} := \begin{bmatrix} B \\ I \end{bmatrix}. \tag{3}$$

This state-vector extension allows us to define the following cost function for the LQR strategy:

$$\sum_{k=0}^{\infty} \hat{x}_k^T Q \hat{x}_k + r\hat{\delta}_k^2, \tag{4}$$

where the matrix $Q > 0$ and $r > 0$ are design choices. The LQR gain K is computed such that the state-feedback law $\hat{\delta}(\hat{x}) = K\hat{x}$ minimizes the cost function Eq. (4), subject to the unconstrained discrete-time dynamics $\hat{x}_{k+1} = \hat{A}\hat{x}_k + \hat{B}\hat{\delta}_k$.

Once we obtained the matrices for the dynamic discrete time system from lake tests, we can tune Q and r, both via numerical simulation and via field tests. Let us hence choose the numerical values $Q = \text{diag}(1, 3, 1)$, $r = 35$, which fit best our requirement of a fast, smooth and overshoot free tack.

A comparison between the dedicated and the LQR tack is depicted in Fig. 3. Although in the first phase (1–2 s) the rudder command from the dedicated controller was more aggressive, the rudder command from the LQR stayed longer close to the saturation than the dedicated controller in the second phase (2–5 s). This difference allows the LQR regulator to execute a faster maneuver, without overshoot.

3 Path Planning

Guiding an autonomously moving vehicle from its current position to some target is referred as *path planning* or *navigation*. In most of the literature, path planning algorithms were mainly designed for wheeled vehicles [12]. The main differences to a sailboat are the lack of the ability of stopping immediately in front of an obstacle or move in every direction. The minimum functional requirements and optimality criteria for a path planning algorithm applicable to a sailboat are:

- sail fully autonomously to a target point (waypoint navigation); optimize speed of the boat and account for the "no-sail zone";
- avoid static and predefined obstacles;
- tactical considerations in order to remain competitive in a regatta.

Notice that in most model sailboats, including Aeolus, the real-time embedded computational capabilities are significantly limited.

In literature, a potential field approach [13] or an approach that includes obstacles as additional no-sail zone in the polar diagram [14] were successfully used on a model sailboat. However, these methods do not include tactical considerations. An alternative to these approaches is presented in [15], where the proposed algorithm covers many of the above mentioned requirements and provides a good basis for extensions. It is assumed that the boat always moves on paths with a constant heading. Therefore, the path planning algorithm projects rays at potential simulated headings $\tilde{\psi}$ into the environment. Each $\tilde{\psi}$ is rated by a weighted sum of several costs, and a low total cost is assigned for feasible directions. A minimum search among the total cost of all simulated directions is then used to find the optimal path in sense of the requirements defined above. This way the onboard computation time is reduced. Further, the projected rays virtually take the whole environment into account, but without simulating future decisions.

The path planning method developed in this works is an extension of the cost function method in [15]. Our extension adds tactical considerations and changes the calculation of the obstacle cost to include the bearing towards obstacles. Further, we introduce a smoothing of the cost function in order to reduce the impact of local minima and increase the safety distance towards obstacles. The inputs are the current heading of the boat ψ, the current position P and the wind direction angle σ. As output, our method supplies the low-level control with $\alpha^\star$ and the maneuver flag. A weighting factor for each cost allows for defining the influence of a particular cost onto the chosen heading. Therefore, we tune the planned trajectory by setting appropriate weighting factors for the individual costs.

The *Target/Wind Cost*, C_w, captures the time to reach the target and takes the speed potential of the boat at different α angles into account:

$$C_w(\tilde{\psi}) = -G_t v(\tilde{\alpha})^\top \cdot T_g(P), \tag{5}$$

where $G_t > 0$ is the weighting constant and $v(\tilde{\alpha})^\top$ is the expected speed of the boat from its polar diagram, which was identified from lake tests. $T_g(P)$ is a vector pointing directly towards the target [15].

For planning paths around obstacles, ideally, the path planner takes into account all obstacles in its environment and searches for an obstacle free trajectory. For

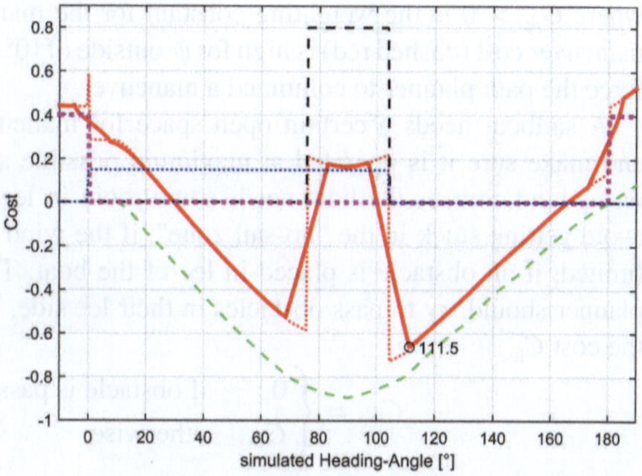

Fig. 4 Cost versus simulated heading for the boat located at start in Fig. 5a and $\psi = 90°$. Total cost (*dotted red*), smoothed total cost (*solid red*), target and wind cost (*dashed green*), obstacle cost (*dashed black*), pass in lee cost (*dotted blue*) and maneuver cost (*dashed magenta*). The minimum cost appears for a reference heading of 111.5°

calculating the *Obstacle Cost* we consider the safety angle $\beta^{\text{safety}} = \tan\left(\frac{r_{\text{obst}} G_{\text{safety}}}{d_{\text{obst}}}\right)$, where r_{obst} is the radius of the obstacle modeled as a circle and d_{obst} is the distance towards the obstacle measured from P. We choose G_{safety} as a factor that increases the obstacle size the safety distance.

If β denotes the bearing to the center of the obstacle, the obstacle cost for the i-th obstacle becomes:

$$C_{o_i} = \begin{cases} G_o & \text{if } \tilde{\psi} \in [\beta - \beta^{\text{safety}}, \beta + \beta^{\text{safety}}], \\ 0 & \text{otherwise} \end{cases} , \tag{6}$$

where $G_o > 0$ is the weighting constant for the obstacle cost, defining the priority for the path optimization. This approach differs from [15] as we use the bearing towards the obstacle, instead of the distance to the obstacle. Therefore, our method allows for passing the obstacle at a closer distance, and that may reduce the length of the path. In Fig. 4, the obstacle cost (dashed black) is relatively high for $\tilde{\psi} \in [84°, 96°]$ as these headings would lead to a collision with the obstacle.

A typical feature of a sailboat is that it slows down with drastic changes in heading. Therefore, maneuvers, especially tacks, should be avoided whenever possible. However, beating against the wind requires tacks. To avoid too many maneuvers, we introduced the *Maneuver Cost*, C_m [15]. It regulates the chance of doing maneuvers on upwind and downwind courses. If no maneuver is necessary, that is when α does not change sign, we assign a low maneuver cost:

$$C_m = \begin{cases} 0 & \text{if } sign(\tilde{\alpha}) = sign(\alpha) \\ G_m & \text{otherwise,} \end{cases} \tag{7}$$

where $G_m > 0$ is the weighting constant for the maneuver cost C_m. In Fig. 4, the maneuver cost (dashed red) is high for $\tilde{\psi}$ outside of $[0°, 180°]$ as these courses would force the path planner to command a maneuver.

A sailboat needs a certain open space for maneuvers to adjust to wind shifts and make sure it is operated at maximum possible speed all the time. Especially on upwind courses the boat needs some room in lee to be able to bear away and avoid getting stuck in the "no-sail zone", if the wind slightly shifts. This ability is limited, if an obstacle is placed in lee of the boat. Therefore, if possible the path planner should try to pass obstacles in their lee side. To achieve this, we introduce the cost C_{lee},

$$C_{\text{lee}} = \begin{cases} 0 & \text{if obstacle is passed in lee} \\ G_{\text{lee}} & \text{otherwise,} \end{cases} \tag{8}$$

where $G_{\text{lee}} > 0$ is the weighting factor. In Fig. 4, C_{lee} (dashed cyan) is high for courses below 98°, because these courses would let the boat pass in windward side, which is not desirable.

Finally, notice that in a regatta condition, the path planner should optimize the trajectory with respect to tactical considerations. The COLREG (International Regulations for Preventing Collisions at Sea) rules state that a boat that sails with wind from port side must give way to a boat with wind from starboard side. Therefore, in a tactical way of thinking, the boat should sail with wind from starboard whenever possible in order to not be forced to give way to other boats. Thus, we also add a tactical cost C_{COLREG} to incentivize sailing with wind from starboard hull.

$$C_{\text{COLREG}} = \begin{cases} 0 & \text{if } \alpha > 0 \text{ (wind from starboard)} \\ G_{\text{COLREG}} & \text{otherwise,} \end{cases} \tag{9}$$

where G_{COLREG} is the weighting factor for the COLREG cost C_{COLREG}.

4 Field Results

We have done extensive numerical simulation of the cost function method using a dynamic model of Aeolus. In Fig. 5a, the boat avoids the obstacle in lee and heads directly towards the target. We also tested the path planner of Aeolus in some lake tests: Fig. 5b shows the logging data, where the path planner was used to guide the boat around a race course with an upwind, downwind and beam reach leg. Two obstacles were placed inside the race field.

(a) (b)

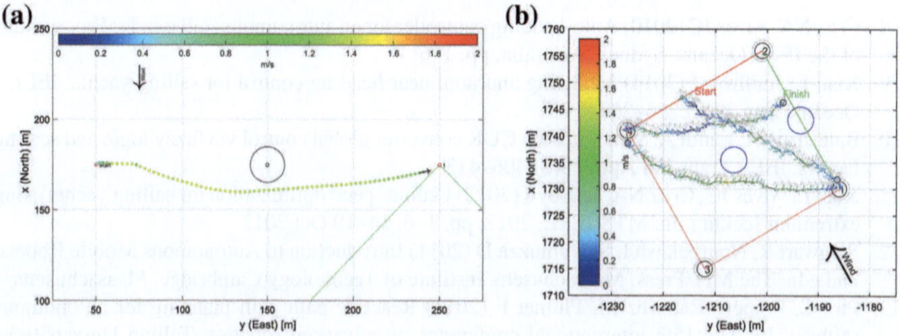

Simulation result. Real lake test result.

Fig. 5 **a** Numerical simulation and **b** field result from a lake test at Lake Zurich, Switzerland. *Arrows* indicate the wind direction and colours are the longitudinal velocity of the boat. The *blue circles* represent obstacles. The boat starts between buoy 2 and 4 and sails upwind to buoy 1, then downwind to buoy 4 and finally beam reach to the finish line

5 Conclusion

We have presented the hardware and software setups of Aeolus, the autonomous model sailboat of ETH Zurich. We have proposed a simple technique to identify a linear state space model of the yaw dynamics relative to rudder commands. This model enabled the tuning of tracking and dedicated maneuver controllers. Real tests on the Lake Zurich, Switzerland, have showed good performance of the low-level control in combination with the high-level path controller. The latter is based on a cost function approach and allows for multi-objective optimization of the boat trajectory. We believe that our setups, design choices and experiments provide useful insights for further research in the field of autonomous sailing.

References

1. AIRMAR. http://www.airmartechnology.com May 2015
2. PIXHAWK. http://pixhawk.org/start May 2015
3. Tranzatto M, Liniger A, Grammatico S, Landi A (2015) The debut of AEOLUS, the autonomous model sailboat of ETH Zurich. In: Oceans, 2015 MTS/IEEE Genoa, 2015
4. Stelzer R, Proll T, John RI (2007) Fuzzy logic control system for autonomous sailboats. In: Proceeding of the IEEE international fuzzy systems conference, pp. 1–6
5. Yeh EC, Bin JC (1992) Fuzzy control for self-steering of a sailboat. In: Proceedings of the international conference on intelligent control and instrumentation, vol. 2, Singapore, pp. 1339–1344
6. Abril J, Salom J, Calvo O (1997) Fuzzy control of a sailboat. Int J Approximate Reasoning 16(3):359–375
7. Emami T, Hartnett RJ (2014) Discrete time robust stability design of PID controllers autonomous sailing vessel application. In: Proceedings of the IEEE American control conference, pp. 1993–1998

8. Cruz NA, Alves JC (2010) Auto-heading controller for an autonomous sailboat. In: Proceedings of the IEEE Oceans, Sydney, Australia, pp. 1–6
9. Xiao L, Jouffroy J (2014) Modeling and nonlinear heading control for sailing yachts. IEEE J Oceanic Eng 39(2):256–268
10. Balestrino A, Landi A, Sani L (2002) CUK converter global control via fuzzy logic and scaling factors. IEEE Trans Ind Appl 38(2):406–413
11. Xiao L, Alves JC, Cruz NA, Jeffroy J (2012) Online speed optimization for sailing yachts using extremum seeking. In: MTS/IEEE, 2012, pp. 1–6, 14–19 Oct 2012
12. Siegwart R, Nourbakhsh I, Scaramuzza D (2011) Introduction to Autonomous Mobile Robots, 2nd edn. The MIT Press, Massachusetts Institute of Technology Cambridge, Massachusetts
13. Ptrs C, Romero-Ramirez M, Plumet F (2011) Reactive path path planning for autonomous sailboat. In: The 15th international conference on advanced robotics, Tallinn University of Technology, 20–23 June 2011
14. Stelzer R, Jafarmadar K, Hassler H, Charwot R (2010) A reactive approach to obstacle avoidance in autonomous sailing. In: 3rd international robotic sailing conference (IRSC) (2010)
15. Guo V, Romero M, Ieng S-H, Plumet F, Benosman R, Gas B (2011) Reactive path planning for autonomous sailboat using an omni-directional camera for obstacle detection. In: Proceedings of the IEEE international conference on mechatronics, Istanbul, Turkey, April 13-15 2011

SWARMON—Real-Time Localization System

Benoit Bourdon, Jean-Jacques Boye, Quentin Descours, Bastien Drouot, Olivier Reynet and Thibault Viravau

Abstract SWARMON is a project that started three years ago at ENSTA Bretagne. The aim is to provide a reliable positioning system to monitor and referee the WRSC. The WRSC is a competition involving small autonomous sailboats (50 cm–5 m), so any additional hardware has to be both small and light. The boats cruise near shore, hence the system can rely on cellular network for data exchanges. The system described in this article is based on both an Android application and a custom hardware tracker on the client side and uses a HTTP RESTful API on the server side. This server allows real-time tracking and replay mode.

1 Introduction

In 2013, a first version of the SWARMON system was developed and deployed during the WRSC2013 in Brest. It aimed at localizing the boats in real time during the WRSC events and to inform the competitors and the public by broadcasting a map on Internet.

This first attempt was based on an Android application running on a waterproofed smartphone and placed on the boat, a simple HTTP server and a dedicated JavaScript-based client to show the map on a screen. Even if this first attempt succeeded in localizing boats, many problems arose due to a lack of system engineering: the GPS coordinates of the boats were sent using SMS, which generated delays, the Android application worked differently depending on the smartphones, the client side application could only be used on one computer at a time and the database could not be accessed via the Internet.

Learning from this first experience, we enhanced the development based on the following requirements:

- The GPS tracking device should be as small as possible to fit in any sailing robot
- The localization process should avoid relying on SMS and use 3G or 4G instead

B. Bourdon · J.-J. Boye · Q. Descours (✉) · B. Drouot · O. Reynet (✉) · T. Viravau
ENSTA Bretagne, 2 Rue François Verny, Brest, France
e-mail: olivier.reynet@ensta-bretagne.fr
e-mail: quentin.descours@ensta-bretagne.org

A. Friebe and F. Haug (eds.), *Robotic Sailing 2015*,
DOI 10.1007/978-3-319-23335-2_9

- The data should be exchanged via Internet through a HTTP front-end
- The client side should only rely on HTTP requests (through a browser, an application or a script)
- The sailing robot should receive a notification when a collision may happen

Therefore it was decided to start a new project in order to:

1. Build a custom hardware tracker to localize the boats
2. Use a web framework to develop the services
3. Create a new Android application to keep using smartphones which are cheap and ubiquitous

The first part of this article explains the technical choices made. The second part describes the SWARMON global system, which consists of a GPS tracker and an Android application interacting with an HTTP server. The third part gives results of the system testing. The last part discusses the possibility of implementing a collision avoidance system.

2 Technical Choices

The key point to keep in mind is that this entire project is the result of successive student projects conducted at the ENSTA Bretagne. Thus, the technical choices were mostly made in regard of their pedagogical potential.

Concerning the development of a device to localize the sailing robots, many solutions have already been developed in the past few years. The development of a custom AIS[1] device (the commercial receivers being too expensive and bulky) has been a considered possibility. However, the students involved were more interested in web development and software engineering than telecommunication. That is why we preferred a web-based approach and chose to base our work on a 3G + GPS module instead.

This web-based approach has also been decided because of the numerous technologies involved. For instance, the use of a web framework allows the iterative creation of a complex system by getting in touch with several concepts and languages.

3 SWARMON Global System

The SWARMON global system is described in the Fig. 1.

The tracking device can be one of the following: a smartphone using our Android application, our custom tracking device based on a 3G + GPS module or any system compliant with our RESTful API. The GPS coordinates (latitude, longitude and GPS timestamp) are gathered and stored on the server to allow later uses. The user can monitor the robots in real time or use the replay mode to display coordinates from

[1] Automatic Identification System.

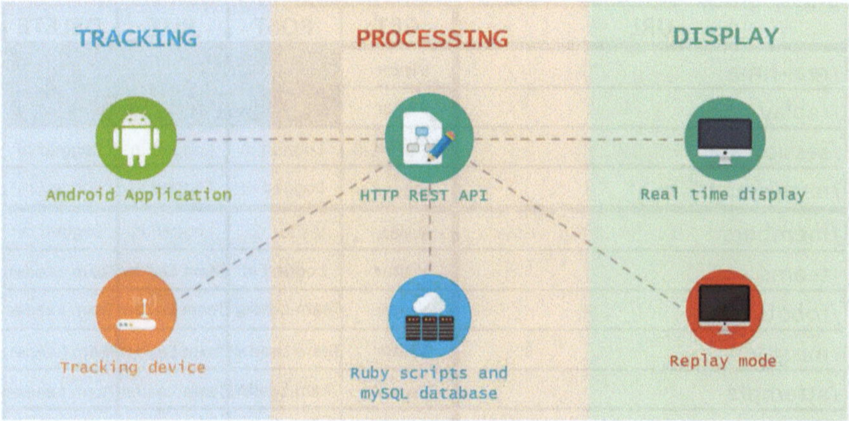

Fig. 1 Description of the SWARMON system

previous events. In both cases, the name of the robot, its team and its category can be displayed directly on the screen with a dedicated colour scheme. The organizers of the competition can create beforehand informative markers which will then be displayed in real time, allowing the public and the teams to identify buoys, obstacles or any area of importance for the current event.

3.1 The Ruby on Rails Web Framework

The server uses Ruby on Rails, an open-source web application framework based on the Model-View-Controller design pattern [1]. This framework has been chosen because it is both an effective and pedagogical tool. By using it, the students have discovered many web standards and technologies but have also gotten in touch with software engineering patterns such as the concepts of 'Convention over Configuration', 'Don't Repeat Yourself' or the 'Active Record Pattern'.

This code and all the hardware description are available on GitHub.[2]

3.2 An Universal and Secured Front End

To interface a GPS tracking system, we chose to develop a HTTP RESTful API [2]. Efficient web services are built using HTTP methods GET, PUT, DELETE and POST [3] and the concept of resources. For instance, in order to push a GPS coordinate into the database, the interaction between the tracker and the server takes

[2]https://github.com/olivierreynet/WRSCMonitor/tree/master.

URL	GET	POST	PUT	DELETE
/real-time	Visitor			
/replay	Visitor			
/session	Logged in	Logged in	Logged in	Logged in
/myaccount	Logged in	Logged in	Logged in	Logged in
/members	Visitor	Visitor	Logged in	Logged in
/teams	Visitor	Logged in	Team Leader	Team Leader
/robots	Visitor	Team Leader	Team Leader	Team Leader
/missions	Visitor	Team Leader	Team Leader	Team Leader
/attempts	Teamate	Team Leader	Team Leader	Team Leader
/trackers	Admin	Admin	Admin	Admin
/coordinates	Teamate	Tracker	Admin	
/what	Android			
/getCurrentMission	Android			
/getMissionInfos	Android			
/gatherCoordsBetweenDates	Android			
/gatherCoordsSince	Android			

Fig. 2 HTTP REST API of the server side, roles and associated permissions

the form of a HTTP POST request on the URL http://www.myserver/coordinates. Another example : if a client wants to get the list of all the sailing robots engaged in the competition, it just has to send a HTTP request using the GET method on the URL http://www.myserver/robots.

The complete HTTP RESTful API is given in Fig. 2. For each HTML request on each URL, the database behaviour can be scripted and the response format can be either HTML or JSON. When using a browser, a simple graphical user interface allows the users to Create, Read, Update and Delete (CRUD) any data, provided that they possess the required permission. As it is described in the Fig. 2, each user of the system has a specific role to which permissions are granted.

To secure the server side, the system implements the HTTP Basic Authentication method to authenticate the users of the system whereas the HTTP Token method is used to authenticate a tracker (hardware or Android Application). Users (referees, sailing teams) have to register and log in to perform some operations (creating a robot, linking a tracker to a robot) while race live display and replay are public. The trackers are authenticated with a token generated by the server which has to be sent by the tracker in each HTTP request.

3.3 Hardware Tracker

The GPS tracking device has been built around the SIMCOM SIM5218E module which provides both GPS and 3G capabilities (Fig. 3).

This module runs a Lua script to retrieve GPS coordinates. It also sends these coordinates to the server and stores them on an external SD card. The module stores the data even if there is no 3G coverage in the current area or if the server is unavailable. The development hardware provided by the maker was useful for the tests, but in order to reduce the size of the tracker, a dedicated interface board has been built, which only embeds the needed features for our use of the module (USB/UART, battery, SIM and SD). Alongside this board and the module, the tracker also includes a GPS antenna, a GSM antenna and a battery (2000 mAh, 3.7 V). The tracker can retrieve up to 10 positions per second, this rate is configured in the Lua script operating the tracker. We choose to use a 1 position per second refresh rate because a higher rate is of little use considering the latence of the 3G network and the relatively low speed of the boats. With this configuration, the tracker achieves an autonomy of 8 h.

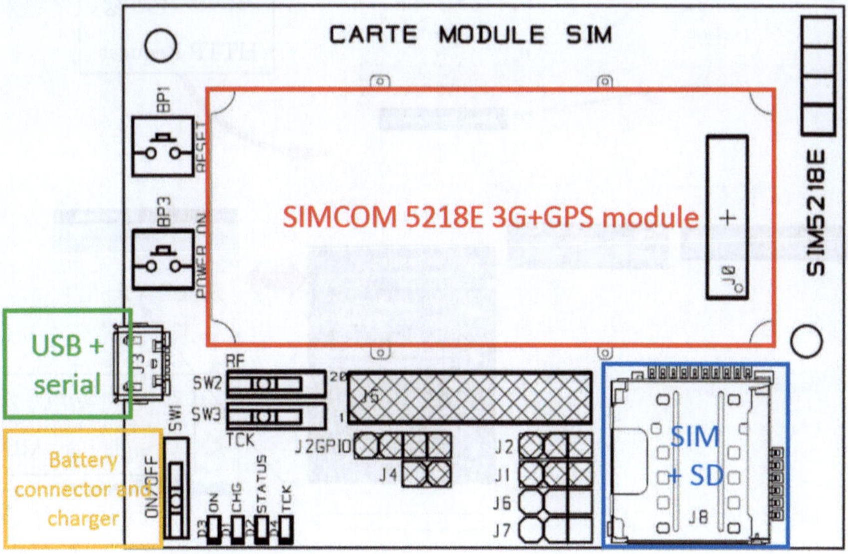

Fig. 3 Electronic description of the SWARMON tracker based on SIMCOM 5218E

3.4 Android Application

A minimalist Android application has been developed to offer an alternative tracking system. This standalone application makes use of the technologies embedded in all modern smartphone to track the sailing robots during the competition : 3G network is used to send the coordinates to the server by using the same API as the tracker. The data are stored in the phone's memory thanks to a local database. In case of a 3G connection loss, the data can be sent to the server afterwards. In Fig. 4, this Android Application is described in terms of activities and processes. Even if the Home button is pressed, the service keeps tracking position and sending messages thanks to a background service.

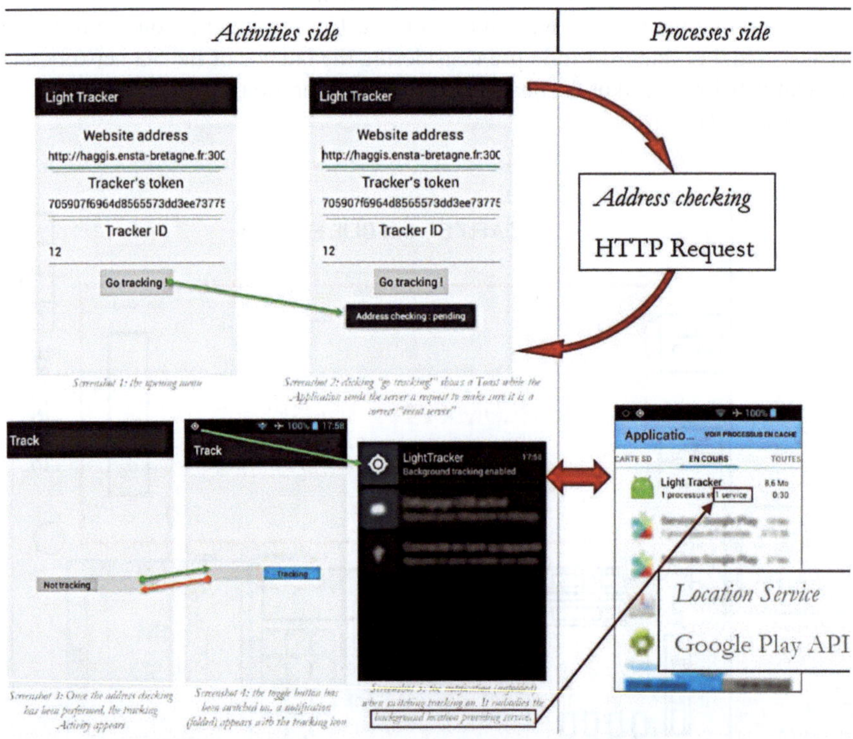

Fig. 4 Description of the Android application in terms of activities and processes

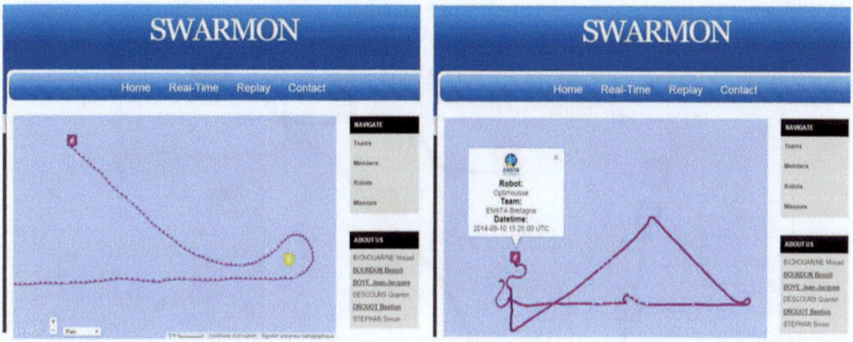

Fig. 5 Real time display (on the *left*) and Replay mode (on the *right*) during the WRSC 2014 in Galway

4 Sea Trials and Testing

In September 2014, an early prototype of the device has been tested on the robots of the ENSTA Bretagne during the WRSC 2014. Even far from the coast of the Galway harbour, the system performed perfectly well and did not suffer from any signal losses. The two modes of display offered by the website at this time are shown in Fig. 5.

In March 2015, tests have been conducted in the Brest harbour (see Fig. 6) to confirm the effectiveness of the GPS tracking device, the Android application and the web application. During these tests, four different Android smartphones and two trackers have been used at the same time to retrieve GPS coordinates, and four devices were accessing the live display application. The server performed as expected by receiving and storing the coordinates in the database. The users were able to follow the testing in real time and in replay mode.

The GPS tracking devices relying on custom hardware also completed the tests and performed better than the smartphones (more accurate, higher refresh rate, more consistent transmission rate). This is due to their two external antennas and their higher reception rate of one position per second (smartphones can retrieve positions every 3 s at best).

To test the Android application we used four models from three different makers.[3] The result is that the application relies heavily on the quality of the smartphone's hardware but also from software optimizations from the maker to improve the autonomy. Even if some models are rather close to the tracking device in terms of accuracy and stability,[4] some others suffer from frequent signal loss or really poor accuracy of the embedded GPS receiver.[5]

[3]Sony Xperia Z1, Wiko Cink Peak, Wiko Ozzy, Nokia X.

[4]Wiko Ozzy.

[5]Wiko Cink Peak.

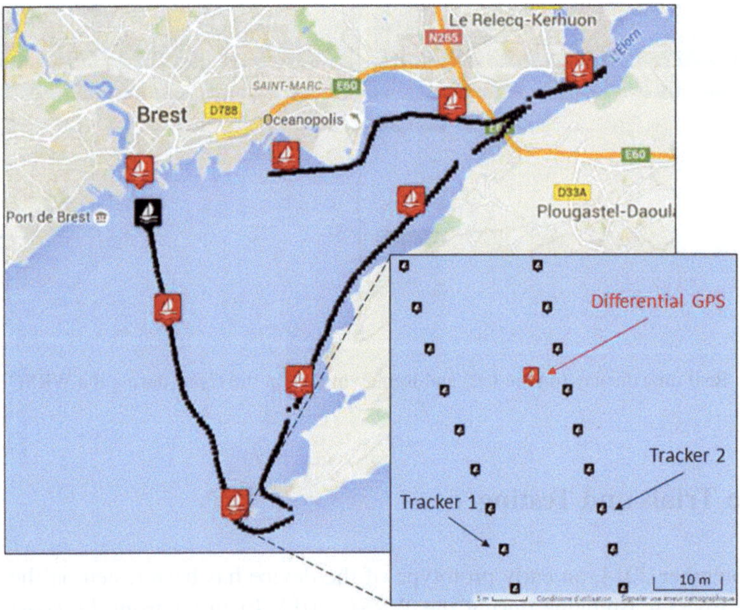

Fig. 6 Geolocalization of two SIMCOM tracking devices (in *black*) compared to a differential GPS (in *red*) during the sea testing

Despite the fact that the accuracy of the GPS position strongly depends on the smartphone, the Android application was stable and completed every implemented features.

5 Possibility of a Collision Avoidance System via HTTP Responses

Now that the tracking system is fully operational, a collision avoidance system can be implemented in SWARMON:

- All the data processing is made in real time on the server side. The latency of the tracking architecture does not exceed a few seconds, which is few compared to the time scale of the sailing robots shiftings.
- The tracking devices provide a serial and/or a USB link which can used to communicate with the sailing robot.
- Each time a tracker posts its localization to the server, the HTTP server responds to the tracker with a HTTP code or a body response[6] to pass information to the tracker.

[6]cf. https://en.wikipedia.org/wiki/List_of_HTTP_status_codes.

Advanced behaviors can then be implemented on the server side or on the sailing robots, thanks to the serial link between the 3G connected tracker and the robots that can be used to exchange data. Rather than just sending an alert, which could lead to another hazardous situation, depending on the drones reaction, the server could describe to each boat the safest route to sail in order to definitely avoid a collision. The server may also suggest the best route to take, depending on the wind, the tides or the ocean currents.

This centralized architecture offers considerable possibilities as already demonstrated by papers from previous WRSC proceedings [4, 5].

6 Conclusion

The SWARMON project is an efficient solution to monitor connected objects. New services as virtual boundaries or swarm behaviours can easily be integrated in this architecture, because all the required tools to compute the data or to communicate with the robots already exist. The only remaining limitation of the system is due to the use of an internet connection which currently restricts its use to coastal environments and prevents any use in open sea. In this configuration, an interesting solution would be to develop a dedicated device based on the AIS system, using the undefined AIS messages to transmit drone and swarm specific information (collaborative behaviours, data exchange, orders).

7 Credits

- N. Brocheton, SWARMON Swarm Monitoring for WRSC 2013, ENSTA Bretagne, 2013
- M. Bichouarine, B. Bourdon, J-J. Boye, Q. Descours, S. Stephan, SWARMON, ENSTA Bretagne, 2014
- B. Bourdon and B. Drouot, Monitor Your Robot: Dedicated monitoring web site for WRSC, ENSTA Bretagne, 2015
- T. Viravau, Android Tracking Application for WRSC, ENSTA Bretagne, 2015
- S. Hunault and T. Zheng, Collision Avoidance System and Autoscoring Capabilities for WRSC, ENSTA Bretagne, 2015

References

1. Ruby S, Thomas D, Hansson DH (2013) Agile web development with rails 4. Pragmatic Bookshelf
2. Richardson L, Amundsen M, Ruby S (2013) RESTful web APIs. O'Reilly Media

3. Gourley D, Totty B, Sayer M, Aggarwal A, Reddy S (2002) HTTP: the definitive guide. O'Reilly Media
4. Ammann N, Biemann R, Hartmann F, Hauft C, Heinecke I, Jauer P, Krger J, Meyer T, Bruder R, Schlaefer A (2010) Towards autonomous one-design sailboat racing: navigation, communication and collision avoidance. In: Proceedings of the 3rd international robotics sailing conference
5. Ammann N, Hartmann F, Jauer P, Krger J, Meyer T, Bruder R, Schlaefer A (2011) Global data storage for collision avoidance in robotic sailboat racing theworld server approach. In: Proceedings of the 4th international robotics sailing conference

Part III
Controllers and Sensors

Autonomous Sailboat Track Following Control

Qian Wang, Mengqi Kang, Jinsong Xu and Jianyun Xu

Abstract A track following controller for autonomous sailboat was developed and applied to a 1.5 m sailboat experiment. There are three modules inside the controller. The module of Local Path Strategy determines the turning process along the specified track. The module of Sail Automatic Control employs the relationship between the sail angle and the apparent wind to achieve the optimal wind drive. The module of Rudder Automatic Control is based on 245 pieces of fuzzy logic rules summarized from the general steering experiences. The lake testing demonstrates the effectiveness of the track following controller in various wind conditions. It has potential for actual application in long-range autonomous sailing.

1 Introduction

As one kind of maritime intelligent transportation vehicles, autonomous sailboats possess the unique advantages in marine data acquisition and monitoring [2], maritime dangerous or illegal behavior surveillance [10]. Using wind as the major power, the autonomous sailboat is self-sufficient in long cruise and continuous operation, which therefore extends the scope of maritime monitoring and saves resources effectively [9].

Extensive research on autonomous sailboat started in the 1990s [1, 2]. There have been some international matches and conferences promoting the development of the

---

Q. Wang · M. Kang (✉) · J. Xu (✉) · J. Xu
State Key Laboratory of Ocean Engineering, Shanghai Jiao Tong University, Shanghai, China
e-mail: gertie.kang@sjtu.edu.cn

J. Xu
e-mail: jinsong@sjtu.edu.cn

Q. Wang
e-mail: xuezixian@sjtu.edu.cn

J. Xu
e-mail: xujianyun101@ sina.com

A. Friebe and F. Haug (eds.), *Robotic Sailing 2015*,
DOI 10.1007/978-3-319-23335-2_10

autonomous sailing boat, such as RelationShip, Microtransat, World Robotic Sailing Championship and International Robotic Sailing Conference [11].

The intelligent control system of autonomous sailboat should have at least three basic functions: global route planning, collision avoidance, and track following control [10]. Global route planning determines the fastest and obstacle-free travel path between the starting and destination points on the basis of ocean environmental information. Collision avoidance is one kind of real-time planning that can divert local course against dynamic obstacles. Furthermore, track following control should ensure the boat to sail along the prescribed path through the automatic control of execution equipment such as sail and rudder. The first two functions place emphasis on the application of decision-making theory, while track following control focuses on application of control theory.

Track following control is a classical problem for all kinds of marine vehicles. However, a large amount of research findings on conventional ships cannot be applied directly to the sailboat which is powered by sail instead of propeller. In the field of autonomous sailboat, sail controller and rudder controller are usually separate and independent: the optimal propulsive force can be obtained through the sail control; course diversion can be achieved by the rudder control.

It is hard to set up a precise physical model for the sail control due to the non-linear dynamic model. An alternative approach is to systemize the artificial process of sail control into the underlying rules [5]. The most simplified rule is the optimum heeling angle corresponding to the maximum sailing speed [4].

The rudder control is not only dependent on sailboat performance, but also related to the environmental factors such as wind, wave and current. The theory of fuzzy logic could integrate the empirical rules into the controller, which is suitable for sailboat rudder control [1, 10].

This research is intended to develop a track following controller involving three modules: Local Path Strategy, Sail Automatic Control, and Rudder Automatic Control. Through the autonomous sailing testing of a 1.5 m sailboat on lake, the track following algorithm is verified. The established fuzzy logic controller only needs to describe linguistically how output variables change with input variables rather than build a dynamic model of the non-linear physical system. It has potential for universal application to different kinds of sailing boats in long-range autonomous sailing.

2 Sailboat Integration and Control Method

2.1 Sailboat Integration

The GRP sailboat with a fabric sail is 1.5 m in length. The main particulars of the hull and the sail are listed in Tables 1 and 2 respectively. The overall configuration, the hull lines and the rudder parameters are presented in Figs. 1, 2 and 3. More design details can be checked from Ref. [12].

Table 1 Main particulars of the hull

The hull data	
Overall length (m)	1.500
Waterline length (m)	1.311
Beam (m)	0.476
Waterline breadth (m)	0.364
Molded depth (m)	0.433
Displacement volume (m^3)	0.015
Draught (m)	0.069
Wetted surface (m^2)	0.493
C_p	0.563

Table 2 Main particulars of the sail

The sail data	
SA (m^2)	1.152
I (m)	2.063
J (m)	0.548
P (m)	1.825
E (m)	0.644
BAS (m)	0.238

The sailing test scenery on the lake is shown in Fig. 4. The sensors equipped in the sailboat include anemoscope, Sail Angle Encoder, AHRS (IG500A of SBG Systems), and DGPS (BDM670 of BDStar Navigation), while the execution system consists of the rudder steering engine and the sail winches. Data communication in Modbus protocol between onboard and onshore computers ensures the real-time data acquisition and control.

2.2 *Track Following Control Algorithm*

The track following control algorithm for autonomous sailboat should achieve at least three functions: local path strategy, sail automatic control, and rudder automatic control.

The most common function of the local path strategy is sailboat course-changing. As shown in Fig. 5, the turning path of three coordinate points can be described as two lines linked with one arc. The radius of gyration is determined by the maneuverability of the sailboat, and the sailboat should be steered before reaching the arc at a distance of about hull length [3]. For this sailboat, the radius of gyration was determined as 8.5 m after sailing trials on the lake.

The rules for sail automatic control are as follows:

Fig. 1 Overall configuration of the sailboat

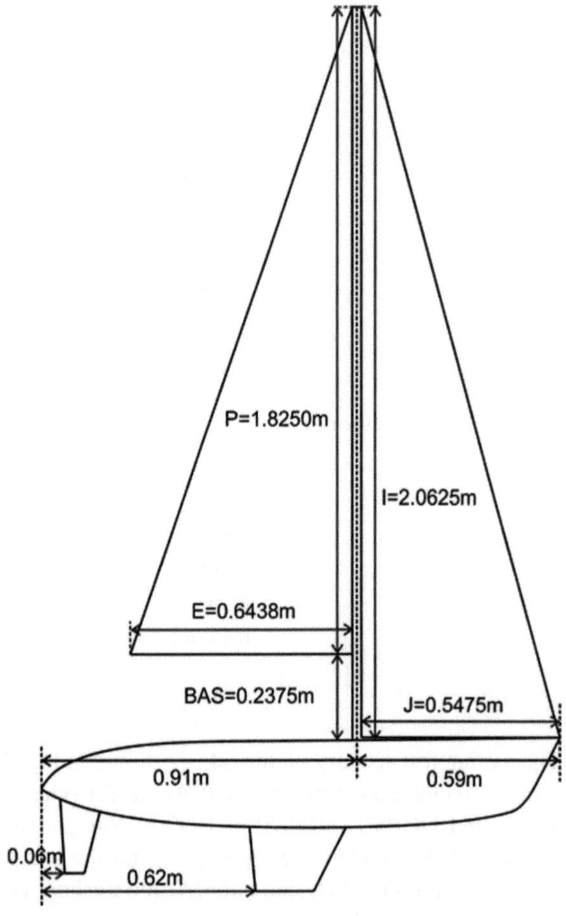

- Sailing against the wind (apparent wind angle: 0° ~ ±30°), tighten the sail to 0°.
- Sailing down the wind (apparent wind angle: ± 160° ~ ±180°), release the sail to 80°.
- Beam wind sailing (apparent wind angle: ±30° ~ ±160°), set the sail angle by linear interpolation.
- Increase the sail angle 15° more when the heeling angle exceeds 30°

Here the maximum sail angle of our sailboat could only reach 80° as a result of the winch restriction.

The automatic rudder control is based on fuzzy logic Mamdani algorithm [7]. There are three input variables including the heading angle error (*e*) between current velocity vector and path direction, the vertical distance (*d*) between current position and specified course, and finally the current yaw rate (*y*). The output variable is rudder angle instruction.

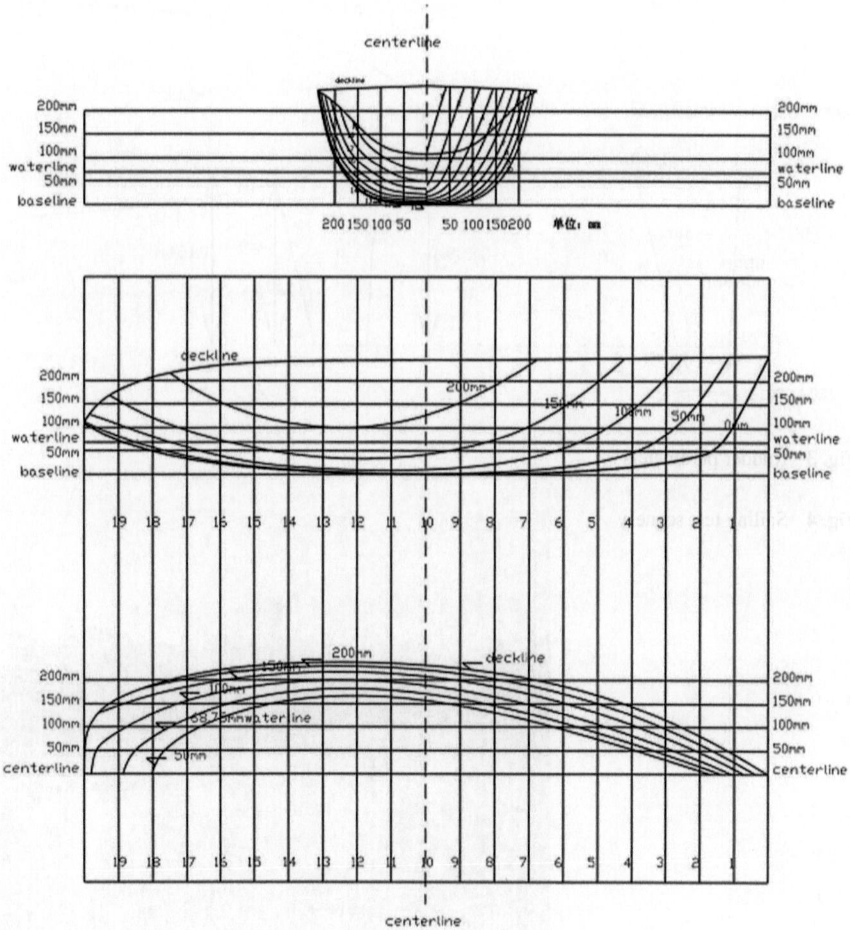

Fig. 2 Hull lines

The fuzzy logic controller involves three procedures: fuzzification, fuzzy logic inference, and defuzzification. As shown in Table 3, each variable is assigned to a linguistically described fuzzy set, which are labeled as: Negative Big (NB), Negative Medium (NM), Negative Small (NS), Zero (ZE), Positive Small (PS), Positive Medium (PM), and Positive Big (PB). The membership functions of each variable are defined in Figs. 6, 7, 8 and 9. The general steering rules are systemized into 245 pieces of fuzzy control rules [6, 8] listed in Tables 4, 5, 6, 7 and 8. The fuzzy logic rules are established according to navigation regulation and the symmetry of rule table. In addition, the integral control item was also added to solve the steady-state error. Physically, it gives the neutral rudder angle for balance.

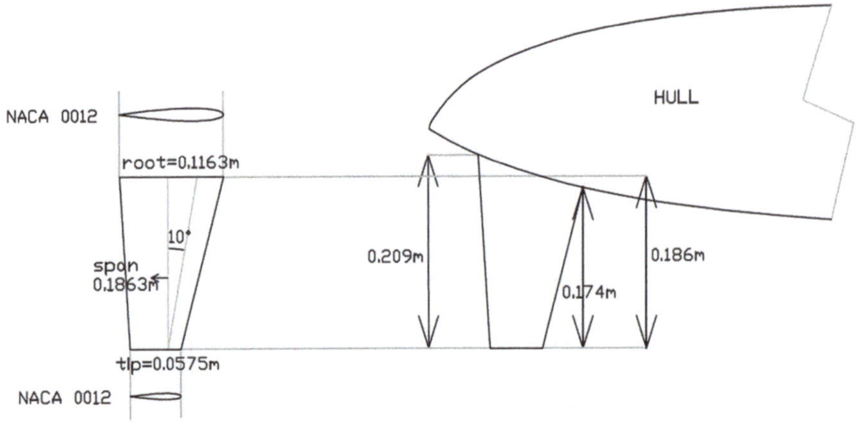

Fig. 3 Rudder parameters

Fig. 4 Sailing test scenery

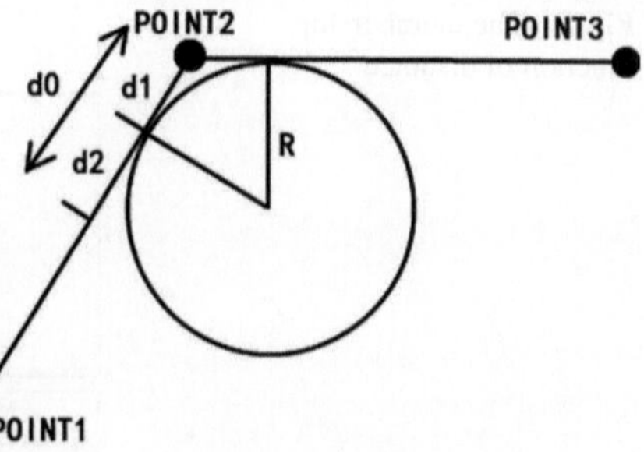

Fig. 5 Sailboat course-changing

Table 3 Fuzzy set of the variables

Items	*Input variables*	*Input variables*
Name	Error	Yaw rate
Fuzzy set category	NB, NM, NS, ZE, PS, PM, PB	NB, NM, NS, ZE, PS, PM, PB
Range of domain	−17° to +17°	−4°/s to +4°/s
Items	*Input variables*	*Output variables*
Name	Distance	Rudder angle
Fuzzy set category	NV, NM, ZE, PM, PB	NB, NM, NS, ZE, PS, PM, PB
Range of domain	−1 to 1(Standardized by d/L)	−45° to +45°

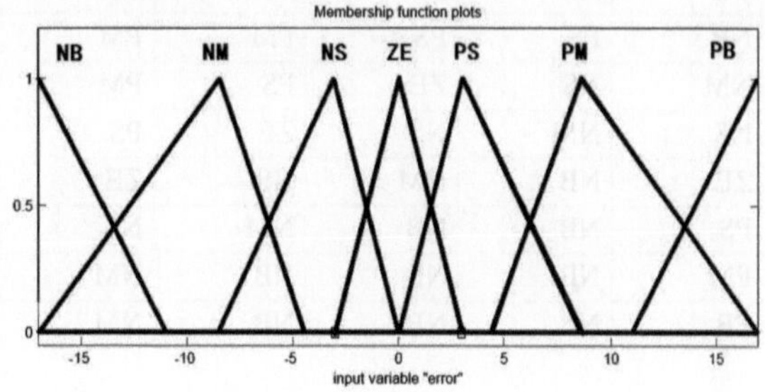

Fig. 6 The membership function of error

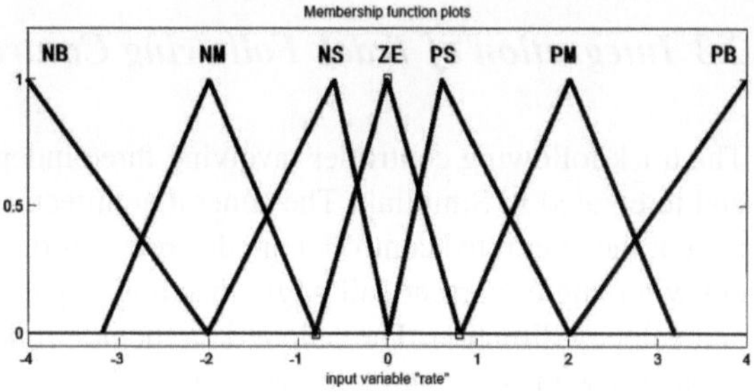

Fig. 7 The membership function of yaw rate

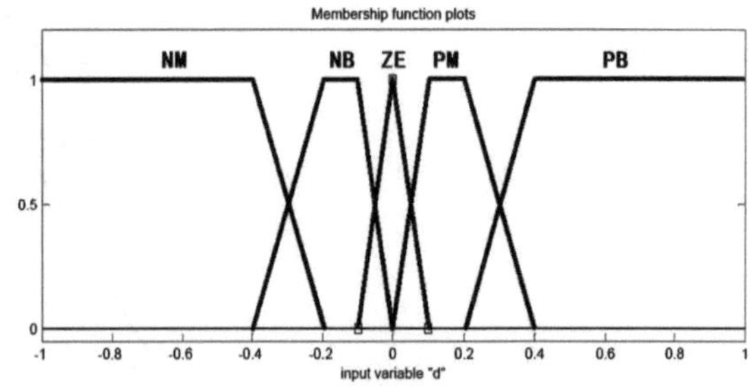

Fig. 8 The membership function of distance

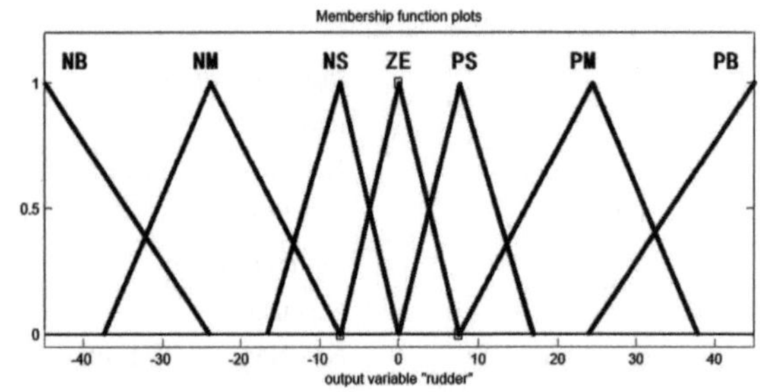

Fig. 9 The membership dunction of rudder

Table 4 The rules when distance is ZE

y\e	NB	NM	NS	ZE	PS	PM	PB
NB	PS	PS	PM	PM	PB	PB	PB
NM	NS	ZE	PS	PM	PB	PB	PB
NS	NM	NS	ZE	PS	PM	PB	PB
ZE	NB	NM	NS	ZE	PS	PM	PM
PS	NB	NB	NM	NS	ZE	PS	PS
PM	NB	NB	NB	NM	NS	ZE	ZE
PB	NB	NB	NB	NM	NM	NS	NS

2.3 Integration of Track Following Controller

The track following controller involving three independent modules was developed and integrated in Simulink. The general architecture is shown in Fig. 10. The input data of the integrated controller are discrete coordinates of the specified path, apparent wind angle, current sail angle, heading angle, heeling angle, current position, and velocity direction. The output data include rudder angle and sail angle instructions. Those program modules are compiled into Dynamic Link Library (DLL) and executed on the platform of NI Veristand in onshore computer. The output real-time rudder angle and sail angle instructions are transferred to the onboard computer to achieve the synchronous control.

Table 5 The rules when distance is NB

y\e	NB	NM	NS	ZE	PS	PM	PB
NB	PS	PS	PS	PS	PM	PB	PB
NM	NM	NS	NS	ZE	PS	PB	PB
NS	NB	NM	NM	NS	ZE	PM	PB
ZE	NB	NB	NB	NM	NS	PS	PB
PS	NB	NB	NB	NB	NM	ZE	PM
PM	NB	NB	NB	NB	NB	NS	PS
PB	NB	NB	NB	NB	NB	NM	NS

Table 6 The rules when distance is NM

y\e	NB	NM	NS	ZE	PS	PM	PB
NB	PS	PS	PS	PM	PM	PB	PB
NM	NS	NS	ZE	PS	PM	PB	PB
NS	NM	NM	NS	ZE	PS	PM	PB
ZE	NB	NB	NM	NS	ZE	PS	PM
PS	NB	NB	NB	NM	NS	ZE	PS
PM	NB	NB	NB	NB	NM	NS	ZE
PB	NB	NB	NB	NB	NM	NM	NS

Table 7 The rules when distance is PB

y\e	NB	NM	NS	ZE	PS	PM	PB
NB	PS	PM	PB	PB	PB	PB	PB
NM	NS	PS	PB	PB	PB	PB	PB
NS	NM	ZE	PM	PB	PB	PB	PB
ZE	NB	NS	PS	PM	PB	PB	PB
PS	NB	NM	ZE	PS	PM	PM	PB
PM	NB	NB	NS	ZE	PS	PS	PM
PB	NB	NB	NM	NS	NS	NS	NS

Table 8 The rules when distance is PM

y\e	NB	NM	NS	ZE	PS	PM	PB
NB	PS	PM	PM	PB	PB	PB	PB
NM	ZE	PS	PM	PB	PB	PB	PB
NS	NS	ZE	PS	PM	PB	PB	PB
ZE	NM	NS	ZE	PS	PM	PB	PB
PS	NB	NM	NS	ZE	PS	PM	PM
PM	NB	NB	NM	NS	ZE	PS	PS
PB	NB	NB	NM	NM	NS	NS	NS

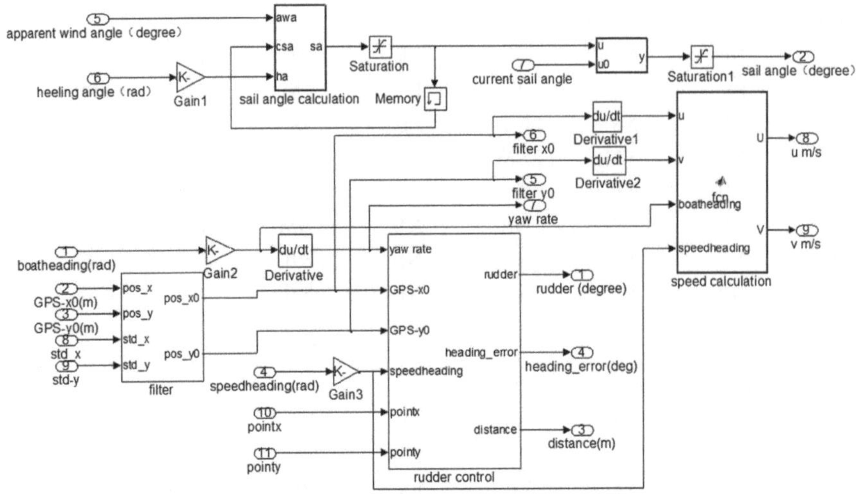

Fig. 10 Architecture of the integrated tack following controller

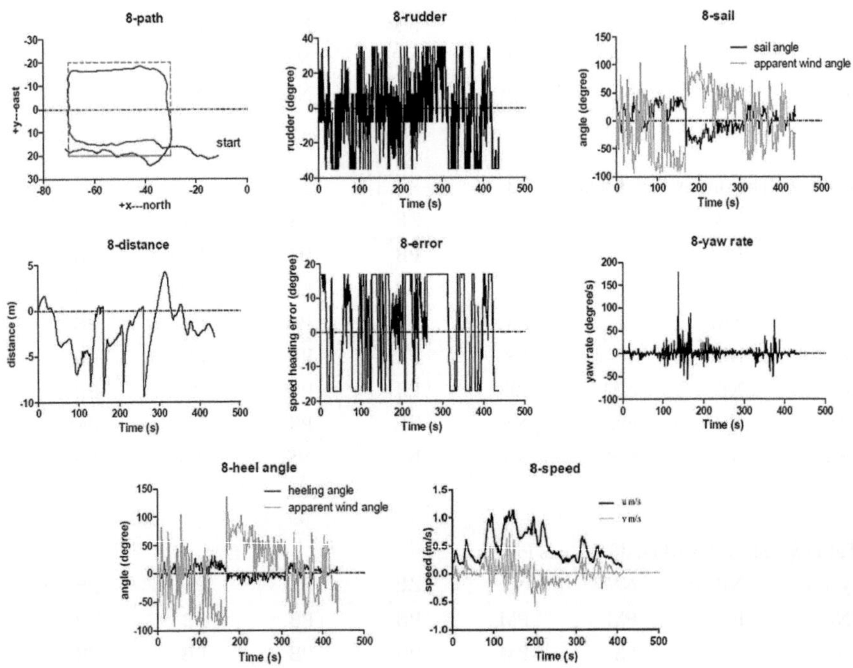

Fig. 11 Test results along rectangular path

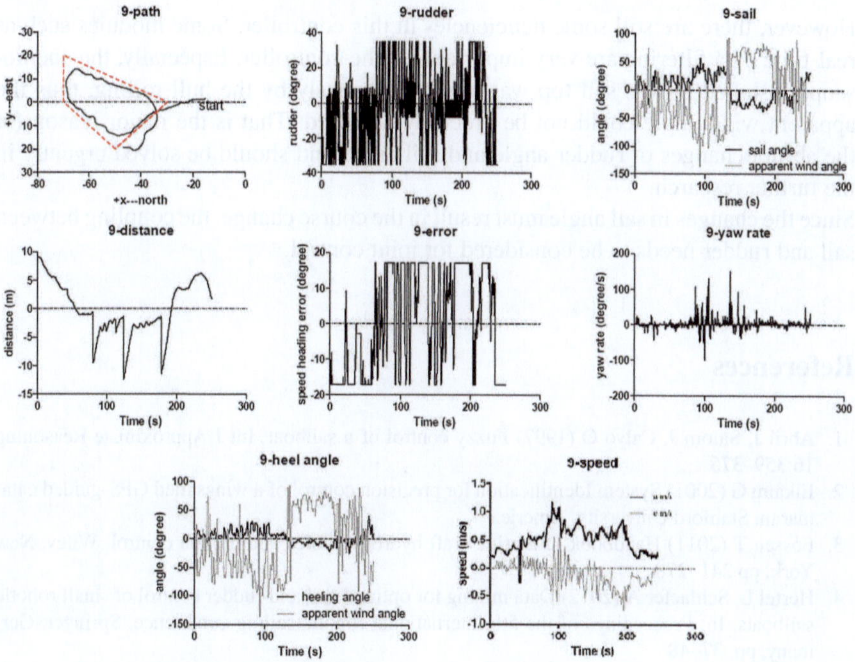

Fig. 12 Test results along trapezoid path

3 Track Following Test Results

The track following test was conducted on the campus lake with the wind speed in the range of 0 ~ 4 m/s. The test results along the rectangular and trapezoid paths are plotted in Figs. 11 and 12 respectively. The eight time-history curves represent the sailing path, rudder angle, sail angle, position distance, heading angle error, yaw rate, heeling angle, and boat speed data in sequence.

4 Conclusions

The autonomous sailing tests on the lake demonstrate the effectiveness of the proposed track following controller in various wind conditions. Comparing with conventional control design, the fuzzy logic controller only needs to describe linguistically how output variables change with input variables without building the non-linear, time-variant dynamic model of the physical system. Thus, the fuzzy logic controller reduces the system complexity and has potential for universal application to different kinds of sailing boats in long-range autonomous sailing.

However, there are still some deficiencies in this controller. Some modules such as real-time data filtering are very important for the controller. Especially, the anemoscope installed on the sail top was affected seriously by the hull rolling, thus the apparent wind angle could not be precisely acquired. That is the major reason for the abrupt changes of rudder angle and sail angle and should be solved urgently in the further research.
Since the changes in sail angle must result in the course change, the coupling between sail and rudder needs to be considered for joint control.

References

1. Abril J, Salom J, Calvo O (1997) Fuzzy control of a sailboat. Int J Approximate Reasoning 16:359–375
2. Elkaim G (2001) System Identification for precision control of a wingsailed GPS-guided catamaran. Stanford University, America
3. Fossen T (2011) Handbook of marine craft hydrodynamics and motion control. Wiley, New York, pp 241–278
4. Hertel L, Schlaefer A (2012) Data mining for optimal sail and rudder control of small robotic sailboats. In: Proceedings of the 5th international robotic sailing conference. Springer, Germany, pp. 37–48
5. Informatik S (2011) Autonomous sailing boats. Salzburg University, Austria
6. Parsons M, Chubb A, Cao Y (1995) An assessment of fuzzy logic vessel path control. Oceanic Eng 20(4):276–284
7. Patyra MJ, Mlynek, DM (2012) Fuzzy logic implementation and applications. Wiley, New York
8. Polkinghorne MN, Roberts GN, Burns RS, Winwood D (1995) The implementation of fixed rulebase fuzzy logic to the control of small surface ships. Control Eng Pract 3(3):321–328
9. Rynne P, Ellenrieder K (2009) Unmanned autonomous sailing: current status and future role in sustained ocean observations. Marine Tech Soc J 43:21–30
10. Stelzer R (2012) Autonomous sailboat navigation. De Montfort University, UK
11. Stelzer R, Jafarmadar K (2011) History and recent developments in robotic sailing. In: Proceedings of the 4th international robotic sailing conference. Springer, Germany, pp 3–23
12. Wang Q (2012) Study on sailing yacht design and display. Shanghai Jiao Tong University, Shanghai

Modeling and Control for an Autonomous Sailboat: A Case Study

Jon Melin, Kjell Dahl and Matias Waller

Abstract A brief review of the literature is used to select a model for sailing boats: The primary purpose of the modeling is to choose a suitable control strategy for autonomous sailing. Also, the model will be used as a state estimator that can handle different, and varying, sampling rates. An experimental setup for evaluating the model and the controller on a Mini 12 equipped for autonomous sailing is described. Practical trials using the Mini 12 are presented.

1 Introduction

Åland Sailing Robots (http://www.sailingrobots.ax/) is a project at the Åland University of Applied Sciences and is concerned with various aspects of autonomous sailing. One main objective is to design a control system that enables autonomous missions with propulsion by sails alone over long periods, from hours to months. The focus is on small sailing vessels between one and four meters in length, with correspondingly limited solar panel and accumulator capacity for supplying electricity to the control and measurement system. A related challenge is power management for the control unit, sensors and actuators. A separate study presents detailed power management solutions for operating different sensors and actuators at different, and varying, sampling rates as well as the microcontroller at different, and varying, clock rates [1] .

J. Melin (✉)
Uppsala University, Uppsala, Sweden
e-mail: Jon.Melin@ha.ax

K. Dahl · M. Waller
Åland University of Applied Sciences, Mariehamn, Finland
e-mail: Kjell.Dahl@ha.ax

M. Waller
e-mail: Matias.Waller@ha.ax

A. Friebe and F. Haug (eds.), *Robotic Sailing 2015*,
DOI 10.1007/978-3-319-23335-2_11

In the present study, a model for the sailboat is evaluated using three criteria:

- How well does the model describe the behaviour of the Mini 12 used for the practical trials?
- Can the model be used to develop a useful control strategy?
- Can the model be used for reliable state estimation?

The reason for selecting these three criteria is connected to the overall strategy for the long-term operation of an autonomous sailing vessel discussed in the above mentioned study [1]. In this paper, the focus is not on a practical implementation of the overall strategy but rather on a flexible solution for evaluating different aspects of the strategy. The setup used for experiments on a Mini 12 sailboat is also described.

2 Model for the Sailboat

A rather exhaustive presentation of ship navigation and control models can be found in the volume written by Fossen [2]. His main focus is on ships and underwater vehicles, but a similar approach to modeling has been extended to sailboats in [7]. The model presented [7] has also been used as a basis for theoretical studies on simulations of and control strategies for sailboats [3, 6]. A simplified model has been presented in [5], with a modest number of parameters and thus attractive for controller design. Given the success of practical trials for autonomous tracking by a sailboat achieved by the team around Jaulin, it seems that the simple model captures the sailboat dynamics essential for controller design. Therefore, the model presented in [5] is also chosen as the basis for the model development of the present study.

The model is given by non-linear differential equations in state space and are provided in Eq. (4). The states $X = \left(x\ y\ \theta\ v\ \omega\right)^T$ are defined in Fig. 1. The model is given in a North-East-Up reference frame, i.e., an easterly x-axis and northerly y-axis. The origin of the boat is its center of gravity and v is the speed in the

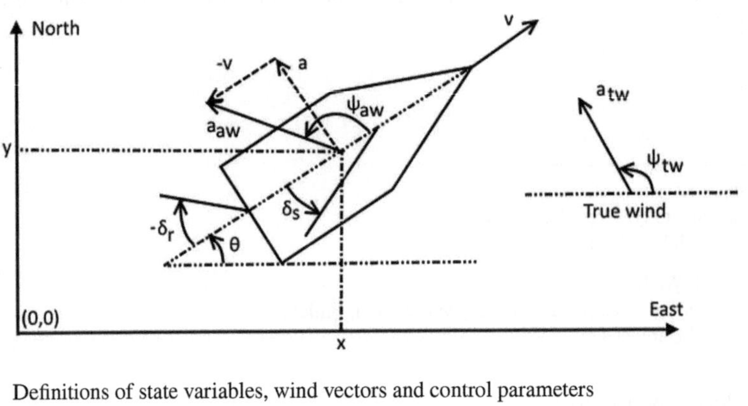

Fig. 1 Definitions of state variables, wind vectors and control parameters

direction of the boat. Wind is described both in reference to the earth and to the sailboat corresponding to true wind (tw) and apparent wind (aw), respectively. Figure 1 also illustrates the angle ψ and speed a of the wind in polar coordinates. For convenience, $W_{p,\text{tw}} = \left(a_{\text{tw}}\ \psi_{\text{tw}}\right)^T$ is introduced. Given the speed and heading of the boat, true wind can be calculated from the apparent wind or vice versa. Apparent wind in Cartesian coordinates relative to the direction of the boat, i.e., the first coordinate corresponding to the heading of the boat, can be calculated from true wind by

$$W_{r,\text{aw}} = \begin{pmatrix} a_{\text{tw}} \cos(\psi_{\text{tw}} - \theta) - v \\ a_{\text{tw}} \sin(\psi_{\text{tw}} - \theta) \end{pmatrix} \tag{1}$$

The corresponding polar coordinates are thus given by

$$W_{p,\text{aw}} = \begin{pmatrix} a_{\text{aw}} \\ \psi_{\text{aw}} \end{pmatrix} = \begin{pmatrix} |W_{r,\text{aw}}| \\ \text{atan2}(W_{r,\text{aw}}) \end{pmatrix} \tag{2}$$

In compact notation, the model is given by

$$\dot{X} = f(X, U, W_{p,\text{tw}}) \tag{3}$$

corresponding to

$$\begin{pmatrix} \dot{x} \\ \dot{y} \\ \dot{\theta} \\ \dot{v} \\ \dot{\omega} \end{pmatrix} = \begin{pmatrix} v\cos(\theta) + p_1 a_{\text{tw}} \cos(\psi_{\text{tw}}) \\ v\sin(\theta) + p_1 a_{\text{tw}} \sin(\psi_{\text{tw}}) \\ \omega \\ (g_s \sin(\delta_s) - g_r p_{11} \sin(\delta_r) - p_2 v^2)/p_9 \\ (g_s(p_6 - p_7 \cos(\delta_s)) - g_r p_8 \cos(\delta_r) - p_3 \omega v)/p_{10} \end{pmatrix} \tag{4}$$

where $U = \left(g_s\ g_r\right)^T$ are the forces generated by sail and rudder. These are given by,

$$\begin{pmatrix} g_s \\ g_r \end{pmatrix} = \begin{pmatrix} p_4 a_{\text{aw}} \sin(\delta_s - \psi_{\text{aw}}) \\ p_5 v^2 \sin(\delta_r) \end{pmatrix} \tag{5}$$

Since the sheet is flexible, the sail cannot hold against the wind and thus stall the boat. This is also accounted for in the model, i.e.,

$$\delta_s = -\text{sgn}(\psi_{\text{aw}}))\min(|\pi - |\psi_{\text{aw}}||, |\delta_s|) \tag{6}$$

where $|\psi_{aw}| \leq \pi$ and sgn is the sign function.

The model is based on traditional translational and rotational inertia affected by forces, yielding changes in position, orientation, speed and rotation. Drift and change in position is described in the first two lines of Eq. (4), due to the speed of the boat and wind interacting with all surfaces except sails, e.g., hull and mast. The acceleration of the sailboat, line 4 in Eq. (4), is affected by three forces: The propulsion from

Table 1 Model parameters for the mini12

p_1	0.03	Drift coefficient
p_2	40 kgs^{-1}	Tangential friction
p_3	6000 kgm	Angular friction
p_4	200 kgs^{-1}	Sail lift
p_5	1500 kgs^{-1}	Rudder lift
p_6	0.5 m	Distance to sail CoE
p_7	0.5 m	Distance to mast
p_8	2 m	Distance to rudder
p_9	300 kg	Mass of boat
p_{10}	400 kgm^2	Moment of inertia
p_{11}	0.2	Rudder break coefficient

the sails, a braking force from the rudder and a tangential friction force, the two latter being proportional to the square of the speed of the sailboat. The rotational acceleration, line 5 in Eq. (4), follows from differences in moments. The forces generated by sail and rudder are approximated by the product of a constant, the velocity of incoming flow (air or water) and sinus of mediums attacking angle, Eq. (5).

Naturally, for a sailboat the propulsion is generated by the sails alone. From a control perspective, this is an interesting characteristic since the traditional perspective defines wind as a disturbance. In this case, however, the disturbance is necessary since without any wind the boat will obviously cease to move, regardless of any efforts made by a controller.

The parameters of the model are determined by comparing simulation results with measured GPS position data as described in Sect. 6. A description of all the parameters, p_i, and the values used in the simulations can be found in Table 1.

3 Control Strategy and Simulations

As a starting point for controller development, a controller inspired by [4] is used. As inputs to the controller, the state variables X, true wind, $W_{p,\text{tw}}$, and a series of way-points, P, are used. The goal is to keep the boat moving along the (linear) trajectories connecting the way-points. The trajectories are assumed to be free of obstacles. Some variables for internal use by the controller are the tack variable, $q = \{-1, 1\}$ used to register the direction of the ongoing tack and a counter k keeping track of current way-points. Also, the controller is provided with tacking angle θ_t, a no-go angle that determines when tacking is necessary. The no-go angle can be defined as a function of wind, but is typically between 30° and 50° on either side of the true wind, ψ_{tw}. In addition, two distance-related parameters are provided to the controller. The distance r, determines the size of the way-point, i.e., at what proximity is the way-point considered to have been reached. The distance, d, determines how close to the desired trajectory the sailboat will keep during tacking. The sheet angle, δ_s, is a linear function of the apparent wind angle, ψ_{aw}.

The control algorithm is as follows.

1. Calculate the distance to next way-point r_1. If $r_1 < r$, the way-point is reached and the way-point counter is updated, $k = k + 1$.
2. Calculate the desired heading θ_r based on the shortest (signed) distance, e, from the boat to the desired trajectory by

$$\theta_r = \beta - 2\frac{\gamma}{\pi}\arctan(\frac{e}{r}) \tag{7}$$

   where β is the angle of the desired trajectory and $\gamma > 0$ is a tuning parameter, i.e., a larger value for γ gives a trajectory of the boat that converges faster to the desired line.
3. Determine mode of sailing, nominal or tack. In nominal mode, go to next step. If tacking is required, that is, true wind lies within the no go zone, q is set to 1 or -1 depending on the direction of the tack and if the sailboat has reached the tacking distance d. The desired heading is correspondingly set: $\theta_r = \psi_{tw} + q\theta_t$
4. Calculate rudder angle, which in the controller is proportional to the sin of the difference between actual heading θ and desired heading θ_r,

$$\delta_r = \text{sgn}(v)\sin(\theta - \theta_r)\delta_{r,\max} \tag{8}$$

   where $\delta_{r,\max}$ is the maximal rudder angle. If the boat is going in the wrong direction, i.e., $\cos(\theta - \theta_r) < 0$, maximal rudder angle is used,

$$\delta_r = \text{sgn}(v)\text{sgn}(\sin(\theta - \theta_r))\delta_{r,\max} \tag{9}$$

5. Calculate sail angle (sheet length), which in the controller is proportional to the angle of the apparent wind,

$$\delta_s = -\text{sgn}(\psi_{aw})\left(\frac{\delta_{s,\min} - \delta_{s,\max}}{\pi}|(\psi_{aw}| + \delta_{s,\max}\right) \tag{10}$$

The closed-loop system was simulated for 600 s using a constant northerly wind of 6 m/s and the results are shown in Fig. 2, the arrow in the middle of the figure indicates true wind direction. The controller was running at 1 Hz. Maximum rudder angle was set to $\pi/6$, and maximum sail angle to $\pi/5.2$. A practical evaluation and the corresponding simulations of the controller are illustrated in Fig. 3. The trial lasted about 13 min, with a western wind of approximately 7 m/s. In the figure, it can be seen that the sailboat oscillated along the trajectory in the beginning of the trial. A possible explanation for this observation is a varying time-delay due to, mainly, communication between the local microcontroller and the external laptop. If a

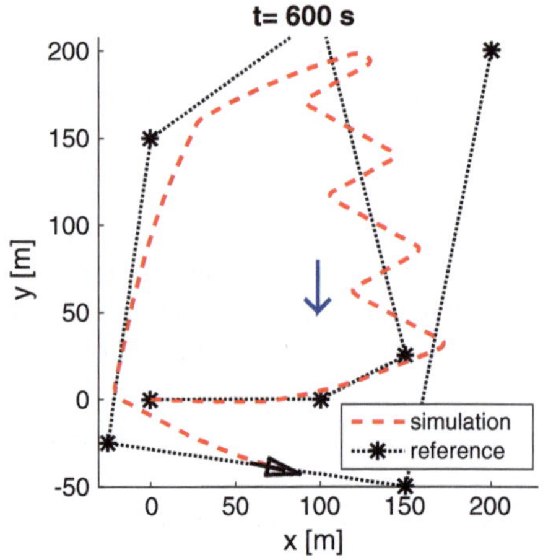

Fig. 2 Simulation over 600 s with model from Eq. (4) and controller from Sect. 3. Desired trajectory, *dotted* (…), simulated path, *dashed* (– –), and wind direction indicated by the *arrow*

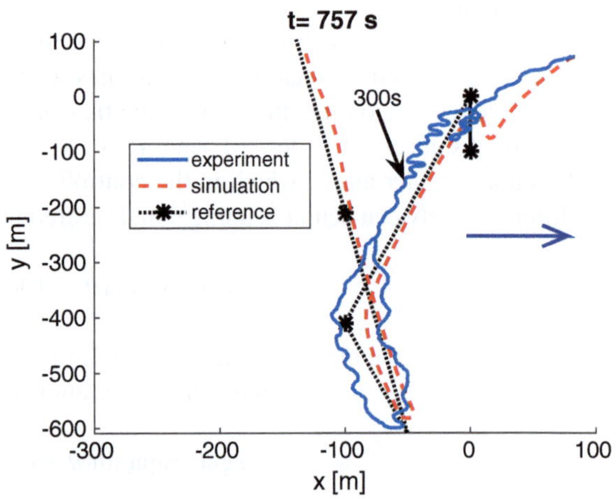

Fig. 3 Experiment with controller compared with simulation. Desired trajectory, *dotted* (…), simulated path, *dashed* (– –), and experiment *solid* (—)

time-delay is included in the model, oscillations can be observed, but further work is required to accurately model the frequencies and amplitudes of the oscillations.

Because of the oscillations, the maximum rudder angle in the controller, $\delta_{r,\max}$ in Eq. (8), was changed from $\pi/6$ to $\pi/9$ after 300 s of the experiment. Although this clearly improved controller performance, the figure illustrates possibilities for further improvement. Also, an obvious drawback of this modification is that maximum rudder can no longer be applied. The trial thus reveals the need to modify the "gain" of the controller without adjusting maximum rudder angle.

In order to explore options for improved control, it can be noted that the control algorithm forms a cascaded control system: The outer control loop, Eq. (7), uses GPS-measurements in order to calculate e, which, in turn, is used by a static nonlinear controller to determine the setpoint θ_r for the inner controller. The inner controller, Eq. (8), uses compass measurements of θ and can be seen as a nonlinear P-controller.

Since the gain of the inner controller cannot easily be adjusted, an obvious possibility is to replace the inner controller by a traditional P-, or PD-controller. At this stage, integral action in the inner controller is not considered for two reasons: First, the outer controller should handle steady-state offsets. Second, for integrating processes, integral action in the controller often significantly degrades performance during step-type setpoint changes.

The use of P- and PD-controllers for the inner controller has been explored in simulations. For a simulated triangle, including a leg of tacking, the controller of Eq. (8) has been compared to a P-controller with a gain of 1 and a digital PD-controller with proportional gain 1, a filter on the derivative term and a derivative gain of 0.5. In addition to the time necessary to complete the triangle, control signal activity given by $\sum|\delta_r|$, control performance given by $\sum|\theta-\theta_r|$ and $\sum|e|$ were calculated. In this comparison, the shortest time is obtained with the P-controller, followed by the PD-controller with the controller of Eq. (8) resulting in the slowest completion of the triangle. Best controller performance for both measures are obtained with the PD-controller, followed by the P-controller and Eq. (8) performing worst. Control signal activity, on the other hand, is lowest for Eq. (8) followed by the P-controller with the PD-controller requiring highest activity. In future work, these indications will be evaluated in practice.

Future work will also consider other possibilities for control. Since different controllers can perform differently depending on, e.g., the desired heading relative to the wind, an option might therefore be to apply different controllers under different circumstances. Other relevant closed-loop characteristics that will be considered are robustness to model uncertainties and noise sensitivity.

4 State Estimation

A desired feature of the autonomous system is very low power consumption during longer missions. As discussed in [1], this can be achieved by shutting down the measurement and control system for longer periods. Informed and safe choices to shut down and wake up the electronic system can be based on reliable estimates of the state of the boat, i.e., mainly the position. Given the form of the model and the promising simulations, future work will explore the possibilities for state estimation with the use of Kalman filter: Can improved estimates of position be obtained at modest computational expense compared to the use of dead reckoning, i.e., constant speed and heading.

A special consideration is that true wind is crucial for the model but only apparent wind can be measured in autonomous sailing. Therefore, a_{tw} and ψ_{tw} need to be calculated from measured apparent wind speed and angle. Also, useful estimates of true wind can be obtained by assuming that wind varies according to a random walk in discrete-time, i.e.,

$$\begin{pmatrix} \hat{a}_{\text{tw}}(k+1) \\ \hat{\psi}_{\text{tw}}(k+1) \end{pmatrix} = \begin{pmatrix} \hat{a}_{\text{tw}}(k) + q_a \\ \hat{\psi}_{\text{tw}}(k) + q_\psi \end{pmatrix} \tag{11}$$

where $\hat{a}$ is used to denote the filtered estimate of a and q is Gaussian noise. En example of measurements and corresponding estimates are illustrated in Figs. 4 and 5. For the same time and approximate location, the Finnish Meteorogical Institue provided a wind direction of 0° and speed of 5–6 m/s, which correspond quite well to the estimates.

In addition, the estimates of true wind in state space need to be compared to measurements of apparent wind. This is achieved with the use of Eq. (1). An interesting possibility is that the model also can be used as an indicator of wind stability—a key element for assessing reliability of position estimates for autonomous sailboats.

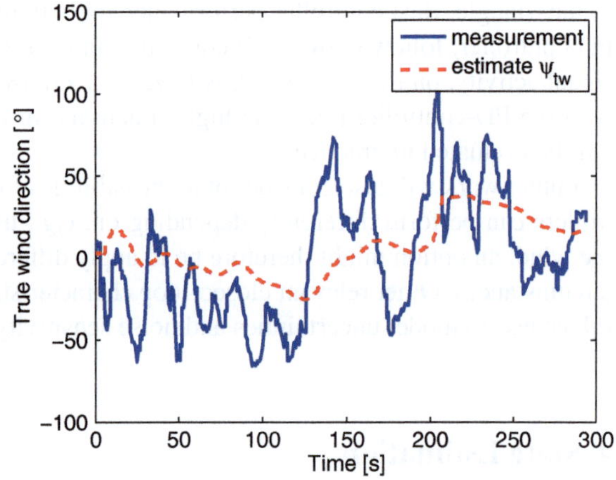

Fig. 4 True wind direction, measured *solid* (—) and estimated *dashed* (- -)

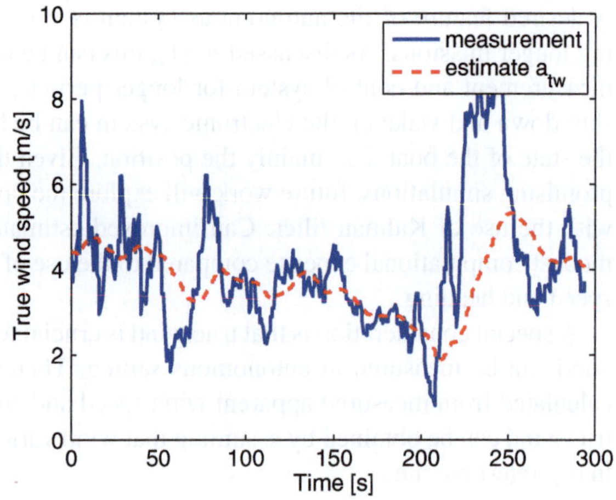

Fig. 5 True wind speed, measured *solid* (—) and estimated *dashed* (- -)

5 Experimental Setup

Although a primary goal is to have an autonomous and energy-efficient measurement and control system, the present experimental setup is designed for easy supervision and real-time evaluation. The design uses a Raspberry Pi (2 model B V1.1) as a local computer in the sailboat. The Raspberry Pi mainly collects and transmits data to an external laptop and controls signals to the actuators. A radio link via XBee XBP24Z7SIT-004 handles communication between the Raspberry Pi and the external laptop. In order to enable rapid controller evaluation in real-time, the laptop uses LabView for supervision as well as to determine control signals. It is also possible to switch between manual and automatic control of the actuators within LabView.

A diagram of the electronic system is provided in Fig. 6. Used sensors are: Wind direction/speed LCJ Capteurs CV7, GPS GlobalSat BU-353 and Compass HMC6343. The sheet and rudder actuators are controlled via PWM signals, the linear actuators are connected via ropes and pulleys to the rudder and sheet. If the XBee link fails, it is, for safety reasons, possible to switch the PWM signals to a traditional radio controller. The range of Xbee is up to 3 Km but the range of the safety system with RC-controller is only about 500 m. It is recommended to not exceed the RC-controller range. The boat, depicted in Fig. 7, used for the experiments is a 4 m long Mini 12 belonging to the 2.4mR-class of the International Sailing Federation. Further development will include measurements of rudder angle via a potentiometer thus enabling accurate rudder control.

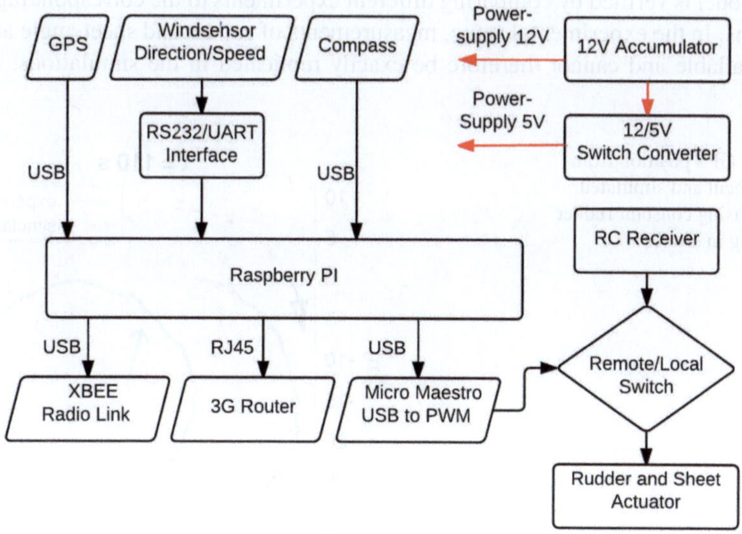

Fig. 6 Electronic system: Components and communication

Fig. 7 The mini 12 during a test

6 Model and Trial Experiments

The model is verified by comparing different experiments to the corresponding simulations. In the experimental setup, measurements of rudder and sheet angle are not yet available and cannot therefore be exactly replicated in the simulations. Apart

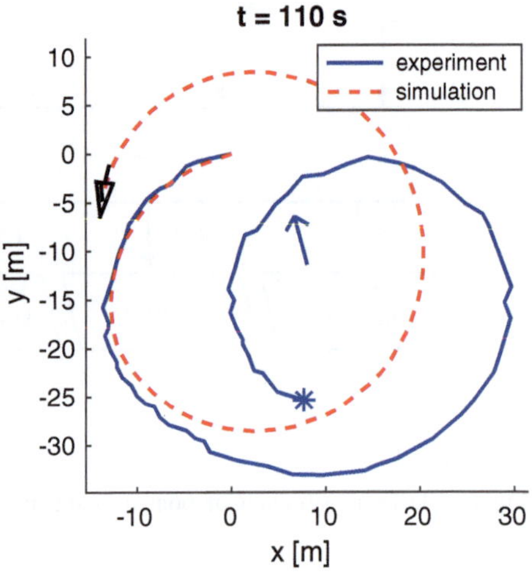

Fig. 8 GPS position from experiment and simulated values using constant rudder resulting in *circles*

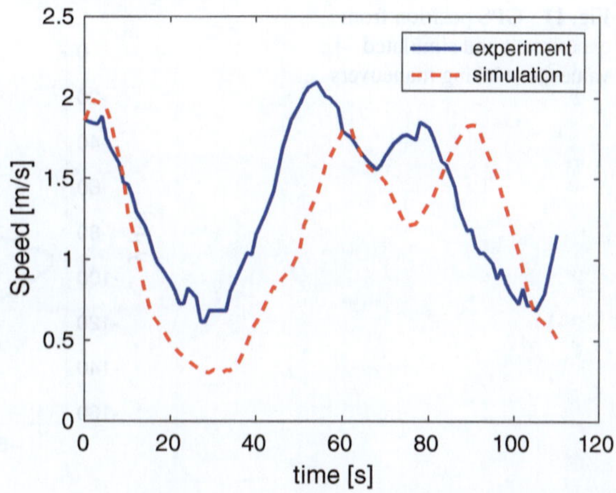

Fig. 9 Boats simulated velocity and measured speed over ground

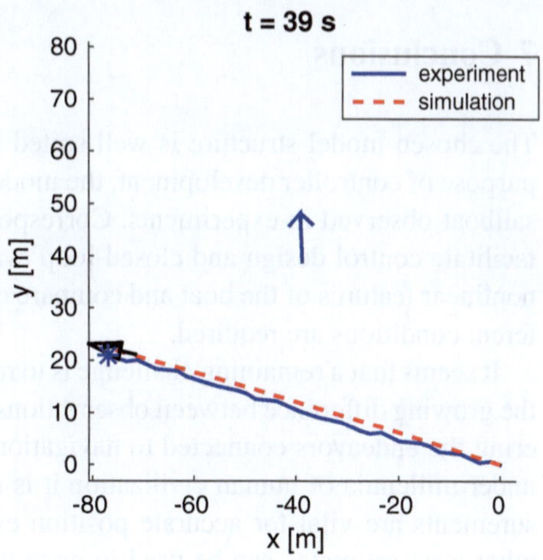

Fig. 10 GPS position from experiment and simulated values for line made with neutral rudder

from measurements of wind speed and direction as well as the initial state of the boat, open simulations are illustrated.

The first test case uses a constant rudder and a maximum sail angle resulting in circles. The ability of the boat, and the model, to turn and handle wind from all directions are thus tested and results are illustrated in Figs. 8 and 9. The second test uses a neutral rudder, constant sail angle and wind from left side. Measurements and simulations are illustrated in Fig. 10. Finally, Fig. 11 shows tacking maneuvers. The simulations uses a neutral rudder angle in between tacking, and maximum rudder during the tacking maneuver. The length of the period with non-neutral rudder is estimated from a graph and corresponding measurements of compass heading. The main sail is trimmed tightly. For these experiments manual control was applied.

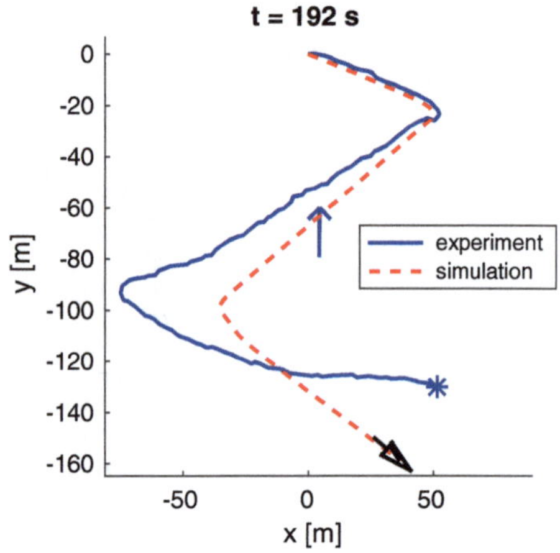

Fig. 11 GPS position from experiment and simulated values for tacking maneuvers

7 Conclusions

The chosen model structure is well suited for simulations and estimation. For the purpose of controller development, the model seems to capture characteristics of the sailboat observed in experiments. Correspondingly, an accurate model can greatly facilitate control design and closed-loop evaluation. In order to further explore the nonlinear features of the boat and compare controllers, more experiments under different conditions are required.

It seems that a remaining challenge is to reliably estimate position as illustrated by the growing difference between observations and simulations in, e.g., Fig. 8. Considering the endeavors connected to navigational satellite systems and the vast efforts under millennia of human civilization it is not, however, surprising that GPS measurements are vital for accurate position estimation. Still, it seems that wind and other state estimates can be used in open water to greatly decrease sampling rates and thus electrical power consumption.

References

1. Dahl K, Bengsén A, Waller M (2014) Power management strategies for an autonomous robotic sailboat. In: Robotic sailing 2014: proceedings of the 7th international robotic sailing conference, pp 47–56. Springer, New York
2. Fossen TI (2002) Marine control systems: guidance, navigation and control of ships. Rigs and underwater vehicles, marine cybernetics
3. Hessberger S (2014) Modeling and simulation of an autonomous sailing boat. Bachelor thesis, ETH Zürich, Switzerland
4. Jaulin L, Le Bars F (2012) A simple controller for line following of sailboats. In: Robotic sailing 2012: proceedings of the 5th international robotic sailing conference, pp 117–129. Springer, New York

5. Jaulin L, Le Bars F (2013) Sailboat as a windmill. In: Robotic sailing 2013: proceedings of the 6th international robotic sailing conference, pp 81–92. Springer, New York
6. Krammer D (2014) Modeling and control of autonomous sailing boats. Master's thesis, ETH Zürich, Switzerland
7. Xiao L, Jouffroy J (2014) Modeling and nonlinear heading control of sailing yachts. IEEE J Oceanic Eng 39(2):256–268

Quantitative Analysis of Various Sail Luffing Sensing Methods

Halie Murray-Davis and Andrew Bennett

Abstract Prior research has shown piezoelectric vibration sensors can detect sail luffing, the acceleration of a flexible sail out of its normal wing state caused by a momentary reversal of the air pressure gradient over the sail. Luffing decreases boat performance by reducing the lift generated by the sail. Yet detecting sail luffing in flexible sails for robotic sailboats is still challenging. This paper presents three methods of sensing characteristics of a luff—air pressure differential which causes the luff, the acceleration of the sail as the luff occurs, and the influence of motion and acceleration of a luffing sail on members placed on the sail. We assess three different sensor types based on cost, ease of use, complexity of electrical interface, power consumption, accuracy of the sensor and amount of noise in sensor readings. To classify the most effective sensor for a given set of constraints, a multifaceted analysis has been performed with a piezoelectric vibration sensor, an acceleration sensor, and a gas pressure sensor. The accuracy and precision of each sensor at sensing sail luffing is evaluated by comparing the sensor output with a plot of the position of a single point on the sail through time generated with computer vision.

1 History & Current Technology

Sails have re-gained popularity as a method of propulsion as the cost of fossil and other incinerated fuels rises and interest in reducing carbon emissions increases. However, traditional, flexible sails require the attention and intervention of a human to keep them trimmed to generate the most propulsive force [8, 13].

Most robotic sailboats approximate the sail setting with respect to the measured wind angle [1, 6, 11]. This is frequently done with trigonometric functions or a simple ratio between sail actuator position and sail position. While effective in many situations, it assumes a uniform system that functions consistently. This is not always

H. Murray-Davis (✉) · A. Bennett (✉)
Franklin W. Olin College of Engineering, Olin Way, Needham, MA 02492, USA
e-mail: halie.murray-davis@students.olin.edu

A. Bennett
e-mail: andrew.bennett@olin.edu

A. Friebe and F. Haug (eds.), *Robotic Sailing 2015*,
DOI 10.1007/978-3-319-23335-2_12

an accurate assumption. For example, if the original sail is replaced with one with slightly different dynamics.

In human operated sailing systems, the operator(s) observes the sail and adjusts it to generate lift. Performance is improved because luffing is avoided [8]. The operator intuitively provides the system with closed loop control. Presently, most robotic sailboats lack this capability [4, 5, 9, 14]. Implementing a method of obtaining feedback about whether the sail is luffing or not can help reduce luffing. It has previously been shown that this feedback can be obtained with use of a piezoelectric vibration sensor [13]. Our work has shown that two additional methods can be used to sense sail luffing: fluid pressure and acceleration sensors. The fluid pressure sensor works by sensing the changes in air pressure which cause luffing. The acceleration sensor senses the acceleration of the sail at the positional extrema encountered by the sail while luffing.

To test the efficacy of these different methods of sensing, representative sensors from each of the three classes of sensor are used. For the piezoelectric vibration sensor, the Minisense 100 [12] by Measurement Specialities was used, see Fig. 1. The analog acceleration sensor ADXL 335 [2], Fig. 2, sensed acceleration. The BMP 180 [3], Fig. 3, was used as a fluid pressure sensor.

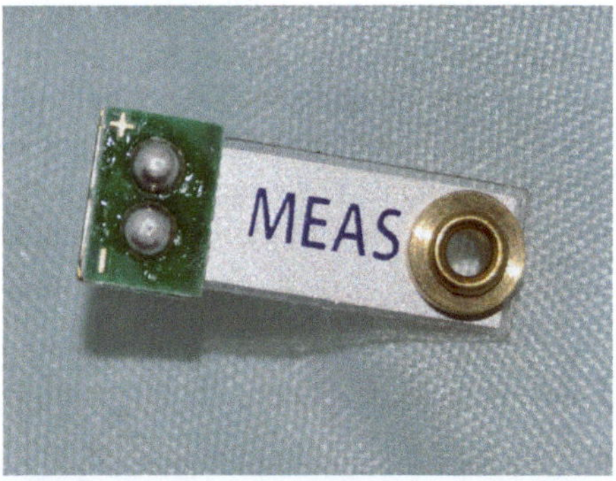

Fig. 1 Minisense100 piezo vibration sensor

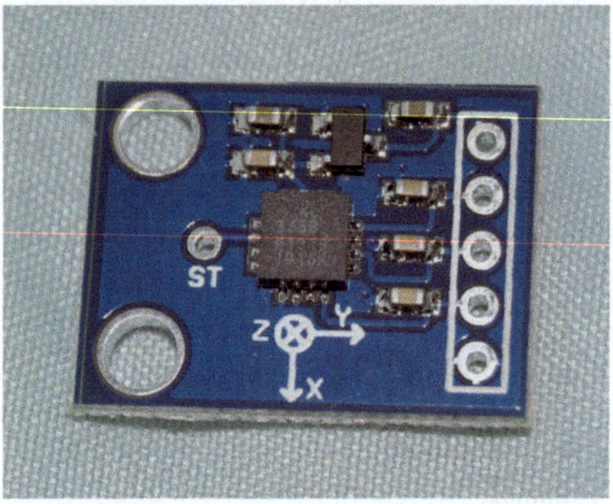

Fig. 2 ADXL 335 analog acceleration sensor

Fig. 3 BMP 180 fluid pressure sensor

Table 1 Economic comparison of the different sensors for both manufacturer and hobbyist vendors

Sensor	Electronics Store	Ebay (China)
Minisense 100	$2.95	$3.98
ADXL 335	$14.95	$2.28
BMP 180	$9.95	$2.46

2 Economic Comparison

All three sensors are affordable when purchased from the less expensive vendors for both small and large scale installation. Cost is recorded in Table 1.

3 Power and Interface Comparison

The Minisense 100 employs a piezoelectric ceramic material to sense vibrations. This produces a voltage as the sensor flexes as a result of the sail luffing; therefore, the sensor does not need to be powered to sense and can have a neutral effect on power consumption. This specific sensor could be replaced by a piezoelectric device designed for power generation and generate power. Only two wires are necessary to read from this sensor–one connecting the sensor to ground and the other detecting the output. A diode is connected in series. The voltages generated on a luffing sail tends to be greater than 0.75 V off the average output of the sensor when the sail is holding shape. Therefore, without accompanying circuitry, the output will saturate the analog to digital conversion offered on most standard computation devices. Placing a diode in series improves this.

The ADXL is powered between ground and either 1.8–3.6 V (standard) or 3.5–5 V, depending on which breakout board is used. The output is an analog voltage proportional to acceleration. The sensor draws approximately 320 μA of current and consumes approximately 1.15 mW of power.

Table 2 A summary of the electrical requirements for the sensors

Sensor	Signal type	GPIO Needed	Supply voltage	Power (μW)	Circuitry
Minisense 100	Analog/generative	Ground, analog input	None	na	Diode
ADXL 335	Analog	Power, Ground, one pin for each axis (min 1, max 3)	1.8–3.6 V	1150	None
BMP 180	I2C	Power, Ground, I2C bus (shared SCL and SDA lines)	1.8–3.6 V	18	Pull-up resistors (2)

The BMP 180 powered between ground and 1.8–3.6 V. It communicates over the I^2C protocol with the data lines also at 3.3 V. It is recommended that pull-up resistors are used with the device, though in most cases the device still functioned without them. The data returned is proportional to pressure and can be converted to pressure. The algorithm for this is given in the data sheet. When reading one pressure value per second, it draws 5 $\frac{\mu A}{s}$ for a power consumption of approximately 18 μW. It uses significantly less power than the ADXL and can be put into an even lower power "sleep" mode to further conserve power.

These features are summarized in Table 2.

4 Data Comparison

In order to compare the effectiveness of the different sensors at detecting sail luffing, the sensors were placed on a sail and tested in an experimental setup.

4.1 Experimental Setup

A sail (luff of 1054 mm) was constructed out of thin BoPET (biaxially-oriented polyethylene terephthalate) polyester film. The sail was designed in SailCut CAD [10], the template's likeness was transferred to the film with marker, then cut by hand with scissors. These panels were constructed into a sail with cellophane tape and vinyl flooring glue. The mast and boom were fabricated from half-inch, balsa wood dowels. The boom vang and downhaul are string. A mounting platform was made with a press-fit hole in a 30 cm x 15 cm x 5 cm wood block. This testing apparatus secured to the top of the car along the center line of the car with four lines attached

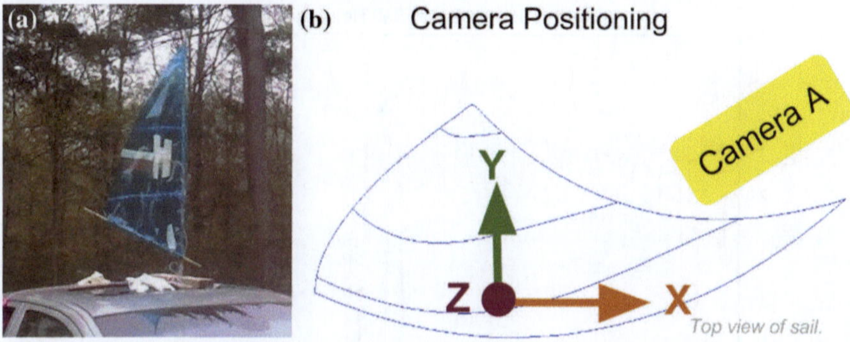

Fig. 4 A picture and schematic diagram describing the experimental setup. **a** Test equipment mounted atop a car. **b** Schematic diagram of positioning of the cameras and reference frames relative to the sail

to the base and tightly tied atop a car roof, see Fig. 4a. The car was driven at 15 miles per hour in a straight line on a low wind night.

Two trial variations were used. In the first, the sail was trimmed correctly so it held its shape and did not luff. In the second, the line securing the sail was slack causing the sail to luff violently in the wind. This simulates an incorrectly trimmed sail or a sail in irons. Five trials of each variation were performed. Finally the sensors were tested in a trial where the sail begins correctly set, but is released midway through the trial so luffing begins. In this trial, we look for a change in sensor output at the transition from holding shape to luffing.

Data from the sensors was collected with an Arduino Uno. The Sparkfun BMP 180 library [15], was used to read data from the fluid pressure sensor. The AnalogRead function from the standard Arduino library was used to read data from the piezoelectric vibration and acceleration sensors.

The data was received from the Arduino over a serial connection to the computer where it was read in by a Python script. When this receives data, it saves the most recent camera frame. Tests were conducted at night with a C270 Logitec webcam directed at a red LED mounted on the sail near the sensors as shown in Fig. 4. A color threshold was then taken on the resulting image and the centroid of the smallest circle then found to track the position of the LED. This image processing was done with OpenCV [7] in real time and was refined after data was collected.

4.2 Data Analysis

In general, sensor output when the sail is luffing is more erratic. This can be seen in the difference in output with characteristic plots of sensor data from trials seen in Figs. 5, 6, and 7 for the sail holding shape and luffing in Figs. 8, 9, and 10. In addition to the magnitude and range of sensor output changing, it also varies more and at a greater frequency.

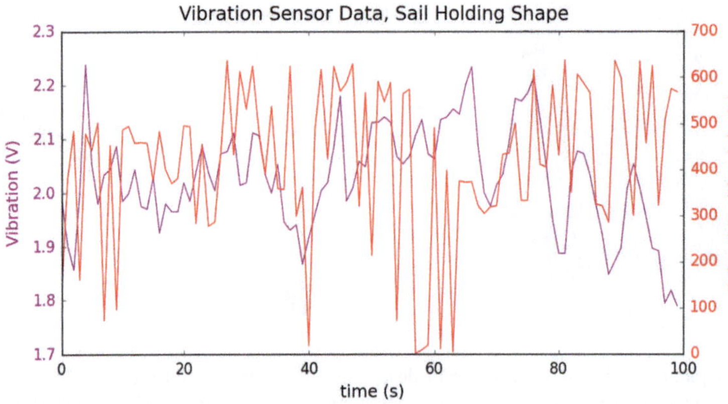

Fig. 5 Output of the Minisense 100 as the sail holds its shape

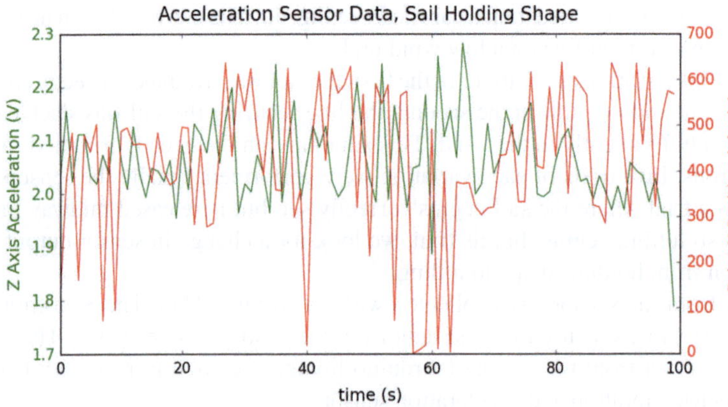

Fig. 6 Output of the ADXL 335 as the sail holds its shape

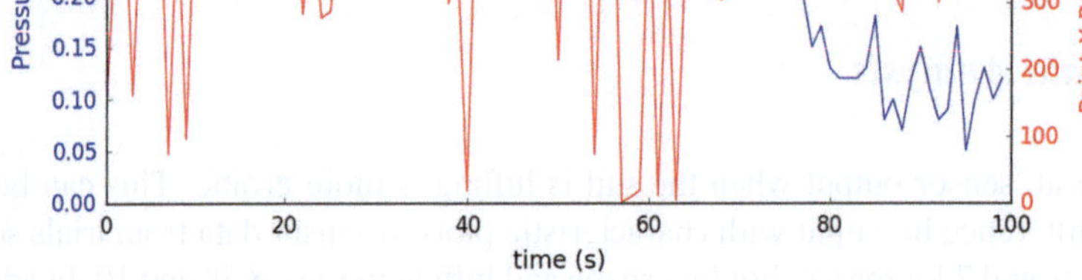

Fig. 7 Output of the BMP 180 as the sail holds its shape

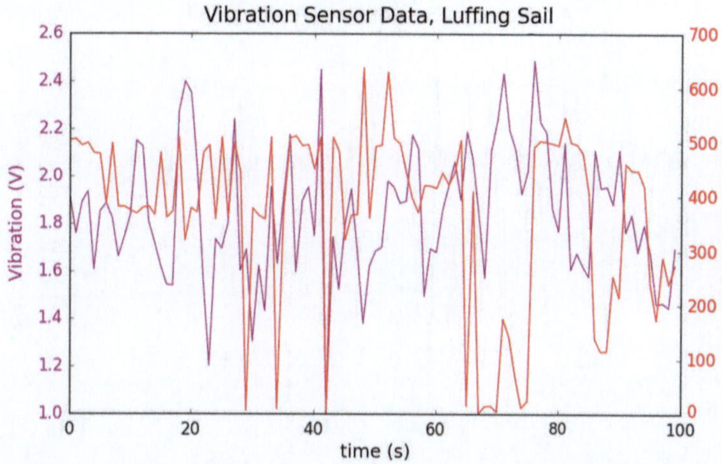

Fig. 8 Output of the Minisense 100 as the sail luffs

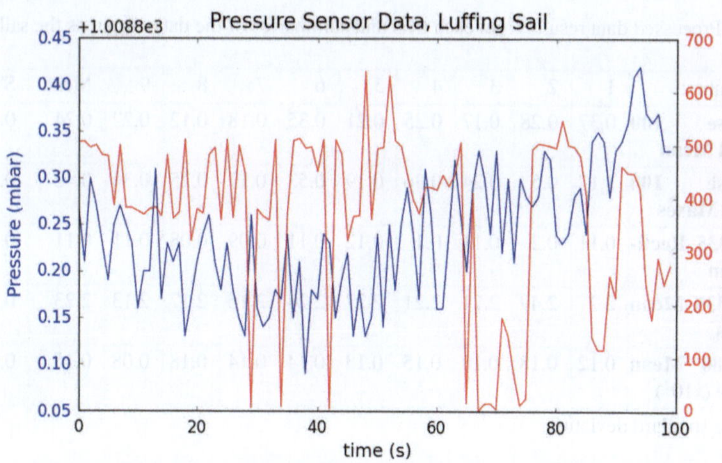

Fig. 9 Output of the BMP 180 as the sail luffs

To obtain a more thorough image of sensor efficacy, we summarized the data. For the Minisense 100 and ADXL 335, we computed the "rectified average" and "average of maxes." The rectified average was obtained by first normalizing the data about zero to account for the sensor offsetting its output around 2 V to avoid outputting negative voltage. To do this, the line of best fit through the data was found with the polyfit function from Numpy. The intercept this yielded was then subtracted from the data to centralize it about zero. Finally, we took the absolute value. This gives us a method of comparing the sensor output for luff events in both directions. To compute the average of maxes, we found the maximum value in a moving time window and then averaged the result. This provides some averaging to account for system noise. The

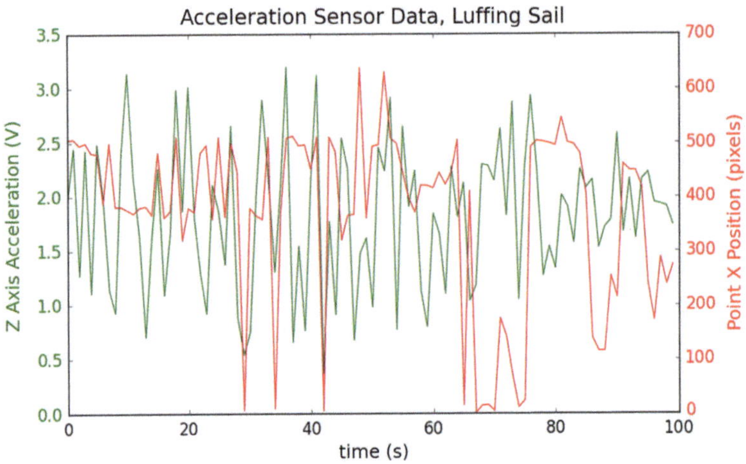

Fig. 10 Output of the ADXL 335 as the sail luffs

Table 3 Processed data results from each trial and summaries of the data output as the sail holds shape

Trial number:	1	2	3	4	5	6	7	8	9	Mean	STD
Minisense 100 Rectified Mean	0.37	0.28	0.17	0.25	0.21	0.35	0.18	0.12	0.22	0.24	0.08
Minisense 100 Mean of Maxes	1.17	0.5	0.24	0.36	0.19	0.53	0.17	0.25	0.34	0.42	0.29
ADXL 335 Rectified Mean	0.11	0.2	0.11	0.1	0.12	0.11	0.09	0.08	0.11	0.11	0.03
ADXL 335 Mean of Maxes	2.2	2.49	2.21	2.21	2.27	2.24	2.16	2.17	2.13	2.23	0.10
BMP 180 Mean Variance ($x10^2$)	0.12	0.18	0.18	0.15	0.13	0.14	0.14	0.18	0.08	0.15	0.03

STD is the standard deviation

average of maxes amplifies the extreme behavior which characterizes luffing. This can be seen in Tables 3 and 4.

The output of the BMP 180 is a bit different as it can be quite hard to decipher what it says about the state of the sail. The pressure can vary due to other factors such as changing speed of the vehicle. Therefore, we instead compute the variance of the data within a moving time window with the var function in Numpy. As the sail luffs, the variance increases. We report the mean variance from the moving time window in Tables 3 and 4.

Finally, we present the results of the data processing over time during an example trial where the sail goes from luffing to holding its shape, see Fig. 11.

Table 4 Processed data from trials when the sail was luffing

Trial number:	10	11	12	13	14	15	16	17	18	Mean	STD
Minisense 100 Rectified Average	0.26	0.22	0.36	0.26	0.25	0.36	0.54	0.31	0.28	0.32	0.09
Minisense 100 Average of Maxes	0.54	1.2	0.52	0.73	0.52	0.58	0.97	0.65	0.56	0.70	0.22
ADXL 335 Rectified Average	0.21	0.49	0.5	0.37	0.3	0.44	0.33	0.49	0.33	0.38	0.10
ADXL 335 Average of Maxes	3.3	3.18	3.18	3.16	3.26	3.26	3.16	3.14	2.83	3.16	0.13
BMP 180 Average Variance ($x10^2$)	0.37	0.24	0.27	0.2	0.11	0.27	0.15	0.21	0.18	0.22	0.07

STD is the standard deviation

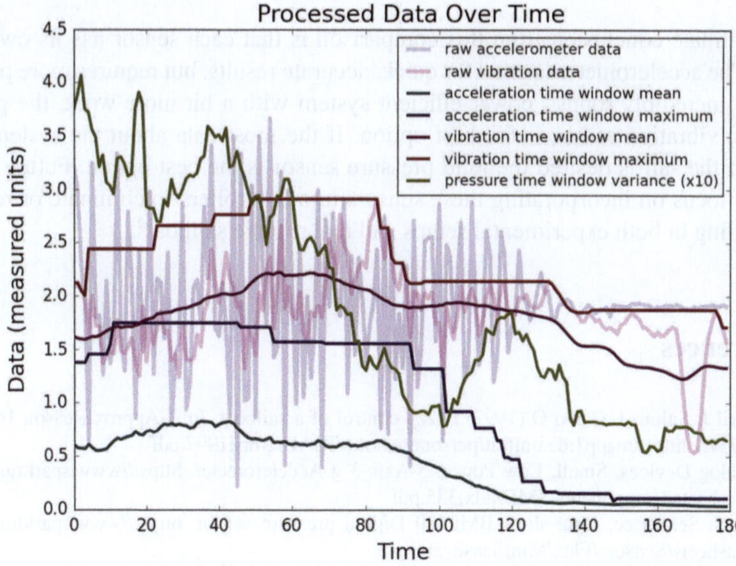

Fig. 11 Output of the processed data over time compared to the raw acceleration and vibration data. As the sail transitions from luffing to holding shape, the output of the processed data decreases, too. Raw acceleration and vibration data is in volts. The pressure variance is scaled by a factor of 10

5 Use Recommendations

As ease of use increases, the specificity of the available data decreases. For the most straight forward, simple method of sensing sail luffing, an acceleration sensor should be used. A threshold can be set above which the sail is considered to be luffing.

The piezoelectric vibration sensor is the next easiest sensor to work with. It excels in situations where minimum power consumption is needed or an extremely robust sensor is desired; however, it outputs significant signal noise. This sensor needs the fewest connection lines and has no electrical components to burn out or short.

The fluid pressure sensor provides a tremendous amount of information about the condition of the sail, beyond if it is just luffing or not. However, it's also harder to distill what this data means about the luffing status of the sail. If additional information about the sail is desired, such as the pressure difference between the two sides of the sail or the pressure in different regions of the sail, for example, to quantitatively test the effect of changing shape and features of the sail on sail performance it can be a reasonable choice. It could be an excellent input for a genetic algorithm to tune sail position or a controller which maximizes the pressure difference between paired sensors.

6 Conclusions and Future Work

The ultimate conclusion from this comparison is that each sensor has its own use case. The accelerometer is good for quick, accurate results, but requires more power. For an incredibly robust, power efficient system with a bit more work, the piezo-electric vibration sensor is the best option. If the most data about the system and state of the sail is desired the fluid pressure sensor is the best option. Future work should focus on incorporating these sensors into controllers to eliminate or reduce sail luffing in both experimental setups and real robotic sailboats.

References

1. Abril J, Salom J, Calvo O (1997) Fuzzy control of a sailboat. Int J Approx Reason 16(3-4): 359375. http://mapp1.de.unifi.it/persone/Allotta/ICAD/Abril1997.pdf
2. Analog Devices, Small, Low Power, 3-Axis 3 g Accelerometer. https://www.sparkfun.com/datasheets/Components/SMD/adxl335.pdf
3. Bosch Sensortec, Data sheet BMP180 Digital pressure sensor. https://www.sparkfun.com/datasheets/Sensors/Flex/MiniSense_100.pdf
4. Dahl K, Bengsn A, Waller M (2015) Power management strategies for an autonomous robotic sailboat. In: Robotic sailing 2014. Springer International Publishing, pp 47–55
5. Eastwell PH: Thunderbird 1: the university of British Columbia's entry in the 5th international robotic sailboat competition. http://www.robosail.org/uploads/2/5/5/8/25586003/development-of-the-ubc-sailbot-final.pdf
6. Giger L, Wismer S, Boehl S, Bsser G, Erckens H, Weber J, Moser P, Schwizer P, Pradalier C, Siegwart R (2009) Design and construction of the autonomous sailing vessel avalon. In: Proceedings of the 2nd international robotic sailing conference, p 1722
7. Itseez, OpenCV. http://opencv.org/
8. Jackson PS, Christie GW (1987) Numerical analysis of three-dimensional elastic membrane wings. AIAA J 25(5):676–682
9. Jnior AG, Arajo AP, Silva MV, Aroca RV, Gonalves LM (2013) N-BOAT: an autonomous robotic sailboat. In Robotics symposium and competition (LARS/LARC), 2013 Latin American, pp 24-29. IEEE
10. Lain R, Lain J: Sailcut CAD. http://www.sailcut.com/
11. Le Bars F, Jaulin L (2012) An experimental validation of a robust controller with the VAIMOS autonomous sailboat. In: Robotic sailing 2012. Springer, pp 73–84

12. Measurement Specialties, Inc., MiniSense 100 vibration sensor datasheet. http://www.adafruit.com/datasheets/BST-BMP180-DS000-09.pdf
13. Murray-Davis H, Barrett D (2015) Piezoelectric vibrational sensor for sail luffing detetection on robotic sailboats. In: Robotic sailing 2014. Springer International Publishing, pp 87–93
14. Schrder C, Hertel L (2014) Development of a low-budget robotic sailboat. In: Robotic sailing 2013. Springer International Publishing, pp 13–24
15. Sparkfun, BMP180_Breakout. https://github.com/sparkfun/BMP180_Breakout

Author Index

A. Friebe and F. Haug (eds.), *Robotic Sailing 2015*,
DOI 10.1007/978-3-319-23335-2

MIX
Papier aus verantwortungsvollen Quellen
Paper from responsible sources
FSC® C105338

If you have any concerns about our products,
you can contact us on
ProductSafety@springernature.com

In case Publisher is established outside the EU,
the EU authorized representative is:
Springer Nature Customer Service Center GmbH
Europaplatz 3, 69115 Heidelberg, Germany

Printed by Libri Plureos GmbH
in Hamburg, Germany